MUSHROOMS
& OTHER FUNGI OF THE MIDCONTINENTAL UNITED STATES

MUSHROOMS
& OTHER FUNGI OF THE MIDCONTINENTAL UNITED STATES

D. M. HUFFMAN, L. H. TIFFANY, AND G. KNAPHUS

IOWA STATE UNIVERSITY PRESS / AMES

DR. DONALD M. HUFFMAN is professor and chairman, Department of Biology, Central College, Pella, Iowa.

DR. LOIS H. TIFFANY is professor, Department of Botany, Iowa State University.

DR. GEORGE KNAPHUS is professor, Department of Botany, Iowa State University.

© 1989 Iowa State University Press, Ames, Iowa 50010

Manufactured in the United States of America

Photographic credits: George Knaphus-Lois Tiffany—2A,B; 4A,B; 20–23; 25–27; 29–41A,B; 42–52; 54–64; 66A,B–78; 80–96; 99–103; 105–6; 108–45; 147–55A,B; 157–73; 175–87; 189–91; 193–214. Donald Huffman—19, 24, 28, 41C, 53, 65, 79, 97–98, 104, 107, 156, 174. Wendell Bragonier—146, 188, 192.

First edition, 1989

Library of Congress Cataloging-in-Publication Data

Huffman, D. M. (Donald M.), 1929–
    Mushrooms & other fungi of the midcontinental United States / D. M. Huffman, L. H. Tiffany, and G. Knaphus. — 1st ed.
        p.   cm.
    Bibliography: p.
    Includes index.
    ISBN 0–8138–1168–6
    1. Mushrooms—Middle West—Identification. 2. Fungi—Middle West—Identification. I. Tiffany, L. H. (Lois Hattery), 1924–   . II. Knaphus, G. (George), 1924–   . III. Title. IV. Title: Mushrooms and other fungi of the midcontinental United States.
QK617.H78   1989
589.2′223′0977—dc19   88-9250
                                                                               CIP

Contents

Preface

I N recent years several factors have combined to renew interest in studies of natural history. Scientific, educational, and recreational groups have come to recognize that organisms have been neglected by our post-Sputnik, molecularly oriented sciences. Our natural habitats and their rich and irreplaceable natural resource of species are endangered by environmental disruptions and threatened with possible extinction. Each group of organisms has its unique appeal, but we think the fungi, especially mushrooms and fleshy fungi, are an unusually diverse and colorful group with distinctive appeal to those with artistic, scientific, photographic, and culinary interests. There has been rapid growth of regional and state mushroom groups, many of which are affiliated with the North American Mycological Association, an organization that synergistically combines the enthusiasm of dedicated amateurs and professional mycologists.

In many parts of the United States, flowering plants, birds, and even geological features have been treated in state or regional field guides, but the lower plants and fungi often are not so well studied or understood. It is our goal to make this guide useful for everyone, but with sufficient depth to meet some of the needs of the more serious scholar. We believe it will appeal to mycology enthusiasts within the larger area of the midwestern prairie states, where woodland and grassland habitats prevail. Midcontinental United States indicates that our studies cover a group of transitional habitats in a region formerly covered in large part by tall grass prairies and woodlands of the eastern deciduous forest and lakes states forest. This region has weather that geographers and meteorologists refer to as midcontinental, and even though these habitat-weather combinations have features in common with adjoining regions, there is a distinctive biological mix.

This is not a complete record of the fungal flora. The midcontinental habitats support a surprisingly abundant variety of mushrooms and other fungi. Our intent is to provide a usable field guide to the regional macroscopic fungi and we have chosen to present about 250 of the more common species. We include a representative sampling of edible mushrooms, some important poisonous species, and other attractive and/or interesting fungi of the region. More information on other fungi is available in the General and Technical References.

A note of **caution**: there are many mushrooms that are not found in this book. If you are confused when attempting to identify a specimen, try one of the other mushroom books. Do **NOT** eat any specimen that cannot be positively identified.

No book of this type is done in isolation, and this is a cooperative venture by many people. We cannot personally thank all who have aided and encouraged us, but several individuals played instrumental roles, and we wish to thank them specifically.

Each author was fortunate to have studied with Dr. Joseph C. Gilman, a mycologist-scholar who was a faculty member at Iowa State University from 1918 until his death in 1966. He was not only extremely knowledgeable in his professional area, but had a broad background of interests and information. His national recognition as a mycologist was evident in his election to offices in the Mycological Society of America for which he served as president in 1945–1946. He influenced many students of mycology who were fortunate to experience his quiet wisdom, sparkling wit, and gentle humanity.

We thank our administrators, colleagues, and students at both Central University of Iowa and Iowa State University who have assisted us, field tested our taxonomic keys, and encouraged us to persist in our publication efforts.

We are also deeply appreciative of the help of Dr. Clark Rogerson, Dr. Walter Sundberg, and Harry Knighton for reviewing the manuscript and providing suggestions for improvements; Dr. Wendell Bragonier for the use of some of his color slides; and Elsie Froeschner for her help with some of the line art.

Finally, we express our gratitude for the financial support of personal and professional friends. In particular, we are grateful to Joan Kuyper Farver and the Stuart Kuyper Memorial Fund, Pella, Iowa, for a gift in honor of the late H. Stuart Kuyper, former president of the Rolscreen Company of Pella, and an enthusiastic supporter of educational and scientific projects. Stu's love for natural history and conservation was evident to those who knew him well, and his generous support of studies in these areas is a tribute to his family, his education, and his vision.

We count ourselves privileged to have known Dr. Joseph Gilman, Stuart Kuyper, Joan Kuyper Farver, and the many others who have aided us and encouraged us, and we dedicate this book to them.

MUSHROOMS
& OTHER FUNGI OF THE MIDCONTINENTAL UNITED STATES

Introduction

FUNGI have been studied and used in all recorded history. The biblical prophet Amos referred to the "blasting of wheat," which was his way of describing the disease that was destroying the wheat, probably a rust disease caused by *Puccinia graminis*. The Vikings were known for their incredible fighting ability and the word berserk in today's common usage refers to the name Berserks, which was given to these crazed fighting Norsemen. Actually there is substantial evidence that they found the mushroom *Amanita muscaria* to be an effective stimulant.

In the middle ages an ailment that caused a gangrenous rotting of the flesh of the feet, hands, nose and ears, resulting in crippling and often death, attacked many people in Europe. St. Anthony's fire, as it was called, was later found to be caused by ergot, a fungus parasite of rye inflorescences. The normal rye grain was replaced by an elongate mass of hard fungal material that was threshed out with the normal grain and then ground with the grain into flour. This fungus contains alkaloids that when eaten may have several effects, such as restricting circulation.

The compounds developed in the ergot bodies cause other reactions. A chemist for a pharmaceutical company was pipetting lysergic acid diethylamide (LSD), which he had isolated from ergot bodies, and sucked some of the compound into his mouth. This was possibly the first "high" on LSD.

Several species of the fungus *Psilocybe* produce hallucinogenic compounds such as psilocybin. These hallucinogens are **illegal** to use, sell, or possess.

Fungi produce another much more widely used and abused drug, ethyl alcohol. Ethyl alcohol is one of the most toxic of the alcohols, but it is the one that the human body can detoxify if it is not imbibed in too large a quantity. One alcohol-producing fungus is the yeast *Saccharomyces cervesiae*; yeasts are also used in baking bread. Allowing the bread to "rise" is simply a way of allowing the yeast to produce carbon dioxide and alcohol. Baking boils off the alcohol, but the spaces that were filled by carbon dioxide and alcohol are left and give the bread its lightness. Omar Khayam spoke of "a loaf of bread, a jug of wine, and thou." Two out of the three is not bad for a tiny fungus.

Fungi are also involved in cheese making. Roquefort is a blue mold cheese, and the fungus *Penicillium roquefortii* is involved in its production. The blue-green material is fungus mycelium and spores, which impart the special flavor. Other cheeses also have fungi contributing to their special flavors and characteristics.

The genus *Penicillium* is also known for the antibiotic penicillin. The original penicillin effects had been noted and used by some doctors, but Sir Alexander Fleming is credited with documenting the effects of the products of *Penicillium notatum*. Many other antibiotics and other medicinal compounds have been produced from fungi. The ergot fungus produces many compounds, some of which are the hallucinogens already mentioned and some of which are used in medicine, including treatment for migraine headaches.

In addition to producing compounds that can help people regain health, there are some fungi that are detrimental to human health. Common ringworm and athlete's foot are among the pesky but not life-threatening fungi that attack humans. Histoplasmosis, a rather common midwestern disease, results from a fungal infection in the lungs of people. In most cases it causes flulike symptoms, the infection site is walled off, and then the victim recovers. In some cases, however, the fungus becomes systemic, causing very serious disease and possibly death. Birds are associated with this disease organism because it grows in the soil in the high-nitrogen condition promoted by droppings. People who work with poultry or who are exposed to dust from soil enriched by bird droppings are most likely to inhale the dry powdery spores. Another fungus with dry, airborne powdery spores is involved with a disease commonly called valley fever. Like histoplasmosis, it involves the lungs and may be encapsulated in the lungs with minor to moderate health problems. If it becomes systemic it may involve other parts of the body, including the brain, and result in very serious disease and often death. This fungus is found in the desert regions of the southwestern United States and adjacent Mexico.

Since fungi are very difficult to treat effectively we can be thankful that they are not our most common disease pathogens. Plants, however, are not as fortunate. Fungi are the most serious pathogens of plants. Plant diseases such as the rusts have the capacity to destroy entire crops. One of the best known incidents of a devastating plant disease is the potato famine in Ireland in 1846–1850. The disease and related economic factors resulted in 1.5 million deaths and massive emigration to the United States, with a resulting drop in population in Ireland of over 50%.

One of the most interesting and important fungal disease situations occurred in 1916–1918. Farmers of the northern wheat-raising regions of the United States were growing wheat with a limited genetic diversity, and in 1916 a very serious outbreak of wheat rust destroyed over one-half of the crop. When the United States entered World War I in April 1917, the 1916 wheat crop was insufficient to meet the needs of America and her allies. In 1917 the wheat crop was again reduced by rust but not to the same extent as the year before. In Germany, however, an epidemic of potato blight (caused by *Phytophthora infestans*) destroyed a major part of the 1917 potato crop. The food problems in Germany were even more pronounced than those in the Allied countries, and eventually Germany succumbed to the military might and greater food supplies of the Allied forces.

Certainly plant breeders are concerned with the major factor of resistance to plant diseases, especially those caused by fungi. Unfortunately, the fungi are able to mutate and produce new and different genetic strains with new potential for successful invasion and destruction; the plant breeder must constantly strive to maintain a diverse genetic base to stay ahead of the relentless fungi.

Sometimes a fungus can live in association with a green plant and not cause serious problems but in fact form a mutually supportive relationship; a mycorrhizal relationship is such a symbiotic association. Many green plants have one or more fungi that are present in the roots, enhancing the uptake of mineral nutrients into the plant. The green plant, however, is the source of food for the growth and life of the fungus. The mycorrhizal fungi often are quite specific, and seedlings from seeds brought from distant lands planted in soil without the required fungus may produce very weak, poor-growing plants; a seedling developing in its native soil will, however, grow very well. Continuing studies of the mycorrhizal phenomenon indicates that probably a majority of plants are involved in such relationships.

We have mentioned various interesting aspects and economic impacts of fungi but have reserved the most important for last. Can you imagine a 10,000-year-old forest where nothing had ever decayed? Obviously it would be impossible because the tangle of fallen trees would be a poor environment for further growth. Even more serious would be the problem of binding the mineral compounds in the trees so that there would be almost no nutrients available in the soil for new growth. Fungi are the major plant-recycling organisms and help replenish the soil nutrients for subsequent generations of plants.

We have spent some time discussing the economic and human-related effects of fungi and have not said much about mushrooms. Actually, the eating of mushrooms has only a minor economic importance compared to the recycling aspects of fungal activity and destruction by plant disease fungi or even possibly to the fermentation of beer. For many of us, however, a bright spring day that dissolves the harsh memories of winter while we search for that elusive first morel also dissolves the humdrum everyday importance of recycling and of plant diseases, and for a short time we think only of edible fungi!

Mushrooms, Mushroomers, and Mycophagists

The fleshy fungi, particularly the mushrooms, are of great interest for a variety of reasons. The possibility of using them as food—Is it edible or poisonous?—is the first concern of many people. For others, the diversity of form and vivid colors are a satisfaction and a delight. There is a challenge in trying to find some kinds of fleshy fungi during their brief season, when they suddenly appear and almost as suddenly disappear. Mushrooms may reappear each year or they may skip a year or more before again producing fruiting bodies. The mysteries of poisons and hallucinogens have created a mushroom folklore containing truths, half-truths, and superstitions.

Groups of organisms have their followers and champions, such as birdwatchers, game-hunters, or fly fishermen; mushrooms are no exception. Consider the following: in the Midwest (and indeed around the world in northern temperate regions) in early May, one may encounter numerous people in the woods, each clutching some sort of bag or basket, gazing intently at the ground, and scuffling in the leaf litter looking for the elusive morel. There is a fascination associated with this vernal ritual that is far out of proportion to the culinary value of the fungus itself. Without dwelling on the eccentric behaviors allegedly assigned to mushroom enthusiasts, it seems safe to say that they are a recognizable cultural variant. After all, people who forage through mud and rain, thorns and bogs, and mosquitos and poison ivy looking for mushrooms and then expressing pleasure in the process are at least as eccentric as ice fishermen, surfers, or spelunkers! If you are a mushroom enthusiast, you are already aware of this unique behavior. If you are not, we invite you to participate. As one author said, "mycologists have more fungi!"

Basic Fungus Biology

Mushrooms are the fruiting bodies of fungi. A typical mushroom consists of a flattened to variously rounded cap, borne umbrellalike at the end of a stalk (1A–F). On the underside of the cap, usually radiating from the stalk, are thin plates of tissue called gills, or lamellae, which are lined on both sides with tiny specialized cells on which the spores, the reproductive units, are produced. The color of the gills of a young mushroom may change from white to another color as the spores mature.

Mature spores are flung from the cell on which they developed and drift in the air until they fall to the ground. If moisture is available, the spores germinate and form threadlike strands of cells called hyphae, which

1 Mushroom cap shapes. **A:** umbonate (with central knob);
B: Companulate; **C:** Plane (flat); **D:** Depressed; **E:** Conic; **F:** Round

form a mycelium, the vegetative body of the fungus (**2**); they must be established where they can obtain food if they are to survive. The hyphae of many mushroom fungi can obtain food and grow in litter consisting of fallen leaves, twigs, and other dead plant parts. Some hyphae will grow only on wood, thus develop in down logs, decaying tree stumps, or in live standing trees.

2 Mycelium in plant debris on ground in a woods. **A:** *Clavaria* sp.; **B:** *Lycoperdon* sp.

A

B

Hyphae may grow for a long time, months or even years, in a particular site (studies indicate that growth can continue for as much as 400 years). Eventually, when environmental conditions (particularly temperature and moisture) are favorable, spore-producing fruiting bodies (such as mushrooms) will be formed. Some mushrooms are most frequently found in rings or arcs. *Chlorophyllum molybdites* and *Marasmius oreades* are probably the most common examples. The most plausible explanation is that after the fungus grows at the center of the ring, the mycelium grows outward in an ever-expanding circle accumulating nutrients. When conditions are favorable, a ring of mushrooms is produced a few inches behind the leading edge of the mycelium. The ring enlarges with each subsequent occurrence.

The hyphae of still other fungi inhabit the living roots of higher plants forming a relationship called a mycorrhiza. These mycorrhizae are very effective in transferring nutrients from the soil into the plant and are often essential to the welfare and survival of both the fungus and the green plant. More information about hyphal growth and nutrients available to fungi is considered in the discussion of mushroom habitats.

A young mushroom, sometimes referred to as an "egg" or "button," develops from an interwoven clump of hyphae (**3A, B**). The cap (pileus), with gills, and the stalk are identifiable even while the entire structure is very small. Some young mushrooms develop with a cover of interwoven hyphae that entirely encloses the young mushroom. As the stalk grows and elongates, the outer cover (the universal veil) breaks, and pieces of it may stick to the cap. Typically, the lower portion encloses the base of the stalk

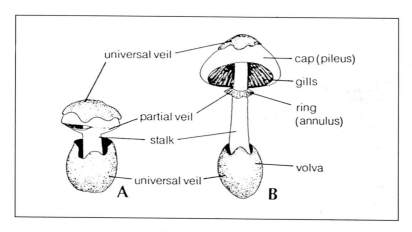

3 Diagram of the principal parts of a complete mushroom.
A: Young fruiting body (button); **B:** Mature fruiting body

as a cuplike structure (volva). Some mushrooms have another cover (partial veil) that extends from the edge of the cap to the stem and encloses the lower surface of the young cap. The manner in which the partial veil breaks when the cap expands is a characteristic feature used in identification. It may persist as a ring of tissue (annulus) on the stalk or, the less common situation, hang as shreds of tissue from the edge of the cap.

As a mushroom matures, the spores (basidiospores) develop on the specialized cells (basidia) on the gills. The gills and spores may remain white or become colored with age; thus, the gills of completely mature mushrooms may appear to be white, pink, rusty brown, yellow-brown, purple-brown, or black. These spore mass colors are used in identification.

To be sure that the color observed is really the color of the spore mass and not the gill color, a "print" should be made to determine the color of the spores. This can be done by cutting off the stalk as close to the cap as possible, then placing the cap with the spore-producing surface down on a sheet of paper for several hours. Cover the cap with a dish or bowl to prevent excessive drying and to avoid wind currents. If the specimen is mature it will release spores, which will accumulate in the gill pattern for gilled mushrooms (**4A**) and in the tube pattern for the boletes (**4B**). Both white and colored spore deposits (prints) will be visible on white paper, but some people prefer to place the cap partly on white paper and partly on colored paper. The prints may be preserved with a light coating of clear lacquer spray, or the paper on which the spores were deposited can be folded with the spores inside for future microscopic study.

The basic macroscopic and field information needed to identify mushrooms is as follows: (1) the presence or absence of a volva and/or annulus; (2) the manner in which gills are or are not attached to the stalk; (3) the mass color of the spore print; (4) the shape, size, color of the cap, etc.; (5) the texture, length, diameter, shape, color of the stalk, etc.; (6) the grouping of the mushrooms, such as single, scattered, clustered, etc.; and (7) the habitat, such as in lawns, on down wood, under coniferous trees, etc.

As you collect fleshy fungi, you will soon notice that many of them do not have the typical mushroom form; they develop other kinds of fruiting bodies with their own macroscopic features. The most common groups can be summarized as follows:

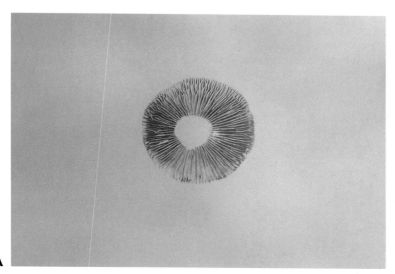

A

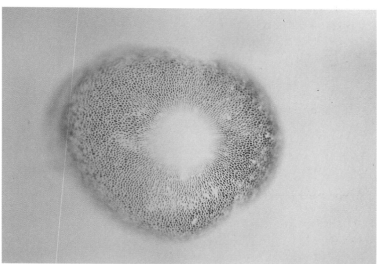

B

4 Spore prints. **A:** Gilled mushroom. **B:** Bolete

A. Groups that fling basidiospores

    1. Boletes (**5**)—have a mushroomlike cap and stalk but have openings or pores on the underside of the cap instead of the typical thin platelike gills of a mushroom. These pores are the open ends of tubes that form the lower surface of the bolete cap, and basidia line the inside of each tube.

Drawing by Elsie Froeshner

5   Bolete

2. Tooth fungi (**6A–C**)—may have various fruiting body forms with cone- or spinelike structures on the underside of the cap. The basidia develop in layers over the surface of each "tooth."

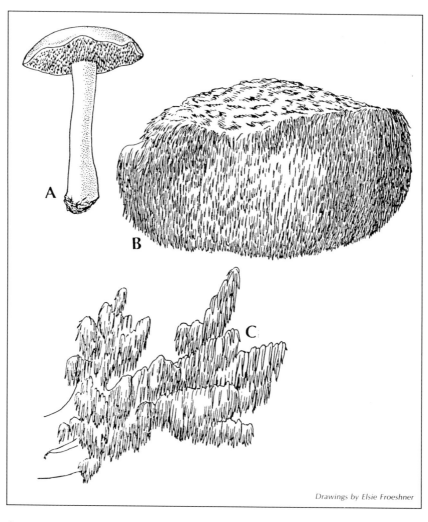

Drawings by Elsie Froeshner

6   Different forms of fruiting bodies of tooth fungi: **A, B, C.**

3. Coralloid or clublike fungi (**7A, B**)—have the same kind of basidia that are produced in layers on the surface of single clubs (**A**) or cover the branches of the fruiting body (**B**).

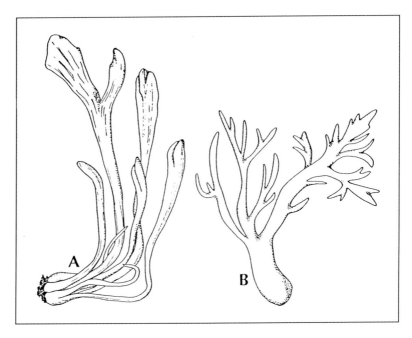

7 Clublike fungi. **A:** Club coral
B: Branched coral

4. Bracket and other shelflike fungi (**8**)—have distinctive, typically non-fleshy bracket or shelflike or flat fruiting bodies. They are common on branches, down logs, stumps, and standing dead or living trees. These fruiting bodies may have an apparently smooth lower surface or a layer of holes similar to those on the lower surface of a bolete cap. The smooth lower surface or the inside of the tubes, whose open ends are the holes, is covered with a layer of basidia. These fruiting bodies are usually leathery, corky, or woody, although a very few are fleshy enough to be edible.

Drawing by Elsie Froeshner

8 Bracket fungi

5. Jelly fungi (**9**)—have firm to gelatinous fruiting bodies that are variously lobed or branched with a cover layer of basidia. Although the basidia are not like those of the previous groups, the basidiospores are flung into the air in the same way.

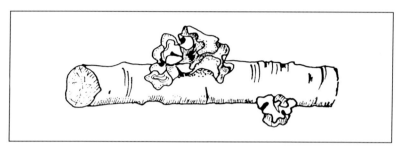

9   Jelly fungi

B. Groups in which basidiospores are not flung

1. Puffballs (**10A–D**)—have dry powdery spore masses that are released by being "puffed" out: the common puffball (**A**) has a hole at the top and a flexible wall, which when depressed releases the spores; the giant puffball (**B**) has an outer layer that breaks away in large irregular pieces, exposing the powdery spores; the stalked puffball (**C**) has a hole at the top of a persistent, flexible cover mounted on a tough stalk; the hard puffball (**D**) has a hard cover that breaks irregularly to release the spores.

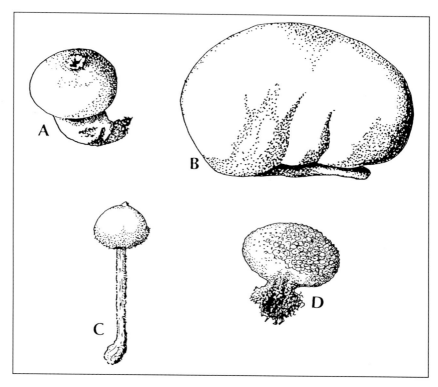

10 Puffball fungi. **A:** Common; **B:** Giant; **C:** Stalked; **D:** Hard

2. Earthstars (**11A,B**)—occur on the ground in the woods or on sandy soils. They look like a puffball with a hole at the top through which the dry spores puff out, but they also have an outer layer (**A**) that folds back in starlike segments (**B**).

11 Earthstar fungi. **A:** Segments closed; **B:** Segments open

3. Stinkhorns (**12**)—have a fleshy hollow stalk with a dark slimy spore mass with a carrionlike odor at the upper end. They are truly distinctive, both in appearance and in aroma!

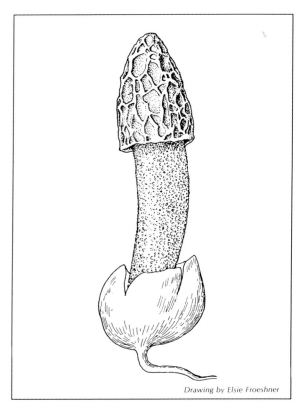

Drawing by Elsie Froeshner

12   Stinkhorn fungus

4. Bird's nest fungi (**13**)—have cup- or goblet-shaped fruiting bodies and usually occur in groups on wood, wood chips, or high-cellulose substrates such as corncobs. Each open bird's nest is 0.5−1 cm in height and contains several lens-shaped "eggs," inside of which the spores are produced.

Drawing by Elsie Froeshner

13   Bird's nest fungi

14   Basidium with four basidiospores

Drawing by Elsie Froeshner

All these types of fungi produce basidiospores (**14**) on special cells that have undergone a particular sequence of events, including meiosis, during development.

Another large group of fungi is characterized by developing spores (ascospores) inside special cells (asci) (**15**), the nuclei of which have had a similar sequence of events during their development. The fruiting bodies of this group of fungi are often small and inconspicuous (less than 1 mm in diameter) except for one group, the cup fungi. Often cup- or saucer-shaped, these fruiting bodies are somewhat fleshy and sometimes brightly colored, the inner surface of which is a layer of asci. The morels, false morels, and saddle fungi are stalked cup fungi, with the pitted or folded upper caplike area developing the typical asci. Some of the more common of these fruiting bodies are as follows:

15   Ascus with eight ascospores

Drawing by Elsie Froeshner

1. Stalked cup fungi (**16A–D**)—have fruiting bodies typical of the morels (**A**), false morels (**B**), saddle fungi (**C**), and earth tongues (**D**). They have a fleshy stalk and a fleshy upper region, which may be spathulate or caplike. The enlarged cap may be smooth, pitted, or folded.

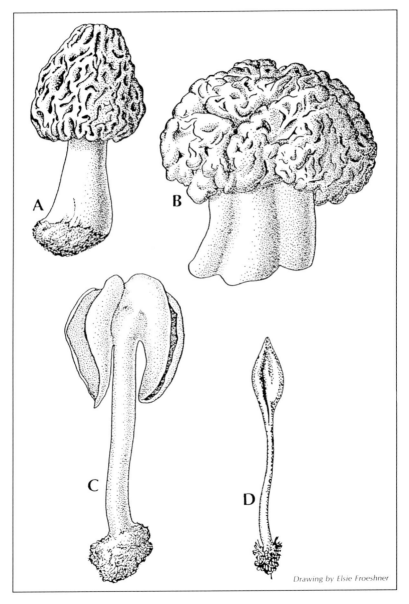

Drawing by Elsie Froeshner

16   Stalked cup fungi.   **A:** Morel; **B:** False morel; **C:** Saddle fungi; **D:** Earth tongue

2. Cup fungi (**17**)—are cup- or saucer-shaped structures that are either stalkless or short-stalked; the colors may be tan, brown, bright red, orange, or yellow. The asci line the inner surface of the cup.

Drawing by Elsie Froeshner

17 Cup fungi

3. Perithecial fungi (**18A, B**)—are more or less flask-shaped fruiting bodies (perithecia), which are normally very small (less than 1 mm in diameter) and barely visible without the magnification of a hand lens or a dissecting microscope. The ascospores may be forcibly ejected from the asci through an opening at the top of the perithecium.

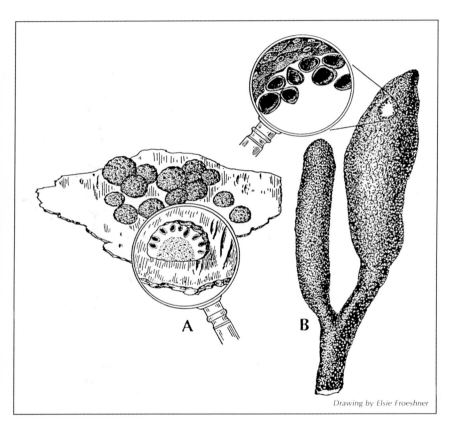

Drawing by Elsie Froeshner

18   A, B: Perithecial fungi; perithecia within stromata

A number of these other fungi are included because some are edible, some are of economic importance in wood decomposition or as plant parasites, and some are simply showy and appealing, such as the slime molds. Our key to the major groups of fungi introduces the groups discussed in this guide.

Naming and Classifying Fungi

Although microscopic examination may be necessary for full identification of some species, generally, good field notes based on careful observation and keys such as those included in this book or other field guides are the tools needed for successful identification. Knowing the season during which a fungus usually develops fruiting bodies also may aid in identification and may be a guide to species expected at a given time in the Midwest.

Not all mushrooms have common names, and some may have several different common names in different countries or in different regions of one country. Obviously, the use of common names can lead to confusion. For this reason a scientific or technical name is necessary. Since the time of Linnaeus (an 18th-century botanist), each organism has been assigned a two-part name, the genus-species *binomial*. To be universally consistent the binomial is written in Latin. While genus-species names initially may seem unfamiliar, they are not difficult to learn, and the species names often are descriptive of the fungus under consideration. For example, *Agaricus campestris* refers to "the *Agaricus* of the meadows," and the common name is the Meadow Mushroom, a clue to its habitat. *Russula virescens* means "the *Russula* that becomes green," a reference to the color of the mature cap. Species names are especially important because most mycological books, including this one, use species epithets as an indexing base. This reinforces familiarity with the species names as the book is used, and the species name is usually retained even if the fungus is placed in a different genus.

The genus-species name is usually followed by the authority (an abbreviation of the name of the person who first described the fungus). For example, *Agaricus campestris* L.: Fr., tells us that Linnaeus was the first to name this fungus, but Fries (in his *Systema Mycologicum* in 1821) used the same name for the mushroom he felt was identical to that described by Linnaeus. The *Systema Mycologicum* serves as the definitive reference for taxonomic treatment of many fungi. As species are redescribed or transferred to other genera by later workers, there may be several changes in authority, and all must be in accordance with the rules of naming set forth in the latest revision of the *International Code of Botanical Nomenclature*. These rules can become complex in application, but this need not concern us here.

Many mycologists currently accept a classification scheme that places fungi in the kingdom Myceteae. In a classification scheme the units are progressively less inclusive; i.e., there may be and usually are several classes in a kingdom, several orders in a class, several families in an order, several genera in a family, and several species in a genus.

Kingdom: Myceteae
Division: Amastigomycota
Class: Basidiomycetes, Ascomycetes, etc. (names end in *mycetes*)
Order: Agaricales, Aphyllophorales, etc. (names end in *ales*)
Family: Agaricaceae, Amanitaceae, etc. (names end in *aceae*)
Genus: *Agaricus, Amanita,* etc. (names are capitalized and often italicized)
Species: *campestris, sylvaticus,* etc. (names are not capitalized and often italicized)

Classification schemes are based on the premise that all species in one genus tend to have characteristics in common with one another in addition to the set of characteristics that is unique to any given species. However, even within a given species there is considerable variation in field characters: genetic changes occur in fungi as they do in other organisms; food, moisture, temperature, and other environmental variables also can affect these characteristics.

Of the many species of higher fungi in the Midwest, we discuss about 250. While we have tried to illustrate and discuss the typical and widespread species, it is inevitable that you will find some that are not included in this book. Several other books may need to be consulted before a particular unusual mushroom can be satisfactorily identified.

Edibility and Toxicity

When collecting wild mushrooms for food, remember that the most common source of information about their edibility has been trial and error; people got sick or died from eating poisonous specimens. Information and evaluations of fleshy fungi as edible, suspect, not recommended, and poisonous have accumulated with time. Thus, identification of a particular species of mushroom becomes very important, though it is not always easy.

Some people believe that all mushrooms are edible, and that all toadstools are poisonous, though it is unclear how this distinction can be made. The problem of mushroom edibility or toxicity is not that simple. The only reliable method of ensuring safe consumption of mushrooms is **accurate**

identification, careful handling and preparation, and tasting of **very small portions** of the mushroom by whoever will ultimately eat that collection. Many fungi contain toxic compounds, the effects of which range from mild or severe digestive upsets to hallucinogenic reactions, to even death. Probably only about 10% of the fleshy fungi are desirable edibles, less than 10% are toxic or poisonous, and the remainder are not normally eaten because of various factors such as bad taste, texture, or small size. Individual sensitivity and allergic responses to specific mushrooms make it impossible to recommend any mushroom as safe for everyone. Learn to accurately identify mushroom species and only eat identified edible specimens that are in good condition. Possibly as many as ⅕ of all mushroom-caused illness results from eating spoiled specimens of otherwise edible species. Avoid mixing collections and exercise your creativity in cooking rather than in exploration of unknown or doubtful species. If not completely certain of an identification, or of the freshness of the mushroom, do **NOT** consider eating it.

Many mushrooms in the Midwest that containing toxins are usually placed in seven groups by toxicologists.

Group I includes mushrooms containing cyclopeptides (i.e., amanitins) that damage kidney and liver tissues; victims have a high mortality rate. The first symptoms occur from 6 to 24 hours following ingestion of the mushroom and include distorted vision, stomach cramps, diarrhea, and vomiting. After these symptoms begin to subside, a far more serious group of events occur, including destruction of liver and then kidney tissue. Death commonly results. In the Midwest *Amanita bisporigera* (quite common), *A. verna,* and *A. virosa* (all deadly) are part of this group. You must learn to recognize the genus *Amanita*. While there are some species that are not toxic, the best policy is **not** to eat **any** species of *Amanita*.

Group II includes mushrooms containing monomethylhydrazine (a component of gyromitrin), which damages liver tissue and has a significant mortality rate. Symptoms occur from ½ to 3 hours after ingestion of the mushroom and include nausea, vomiting, cramps, and diarrhea along with loss of muscular coordination. Mushrooms suspected of containing toxins in this group include species of *Gyromitra*.

Group III includes mushrooms containing coprine, a toxin that induces nausea, a flushed feeling of the skin, chest pains, and vomiting from ½ to several hours after ingestion but apparently only when **alcohol** is in the body or is ingested with the mushroom. *Coprinus atramentarius* is the principle species involved in the Midwest, but *C. quadrifidus* and *C. micaceus* may also produce this reaction.

Group IV includes mushrooms containing muscarine, which, like the species in Group III, affect the autonomic nervous system. Within ½ to 3 hours after ingestion, perspiration, abdominal cramps, salivation, and lacrimation appear, and possibly the reduction of the pulse and a lowering of the blood pressure to critical levels. Death may occur depending on the quantity eaten and the size of the individual; children are more vulnerable than adults. Species in this group include *Amanita muscaria, Inocybe fastigiata, Boletus luridus,* and *Clitocybe dealbata.*

Group V includes mushrooms containing mucimol or ibotinic acid, two closely related toxins that may cause loss of coordination, confusion, dizziness, vomiting, and convulsions within ½ to 3 hours after ingestion. Symptoms are quite variable from person to person, and mortality rates are less than 1%. *Amanita muscaria* and *A. pantherina* are two of the species involved in this group.

Group VI includes mushrooms containing psilocybin or psilocin, two related toxins that may cause hallucinations, headaches, fever, twitching, convulsions, vomiting, and severe dysphoria occurring from ½ to 3 hours following ingestion; mortality rates are less than 1% (one person is reported to have died from eating *Psilocybe cubensis* in 1984–1985). This group of mushrooms is involved with recreational use as well as ritualistic use among some Mexican Indians.

Group VII includes mushrooms that have a wide variety of gastrointestinal toxins that produce symptoms of nausea, vomiting, diarrhea, and headaches within a few hours of ingestion. We do not have mortality rates for this group because of the diverse types of both toxins and mushrooms involved. A sampling of midwestern species involved includes *Boletus luridus, Chlorophyllum molybdites, Clitocybe dealbata, Collybia dryophila, Omphalotus illudens, Ramaria formosa, Russula emetica, Lactarius piperatus,* and *L. torminosus.*

One becomes more and more impressed with the diversity of effects mushrooms may have. Sometimes individuals are seriously affected (possibly an allergic response), while others who eat the same species at the same meal show no effects. The matter of edibility of mushrooms and the possibility of poisoning is very complex. The complexity is multiplied by the following:

1. Individual mushrooms within a species may vary in toxin content.
2. Geographical location may be a variable; i.e., a mushroom species may be toxic in the eastern United States and another variety of the same species collected in the Pacific Northwest apparently can be eaten.
3. Mushrooms reproduce sexually and there may be genetic differences between individual strains, resulting in varying amounts of toxins.
4. The effects produced by toxins vary from person to person.
5. Other foods or substances (such as alcohol) consumed with mushrooms may enhance toxicity.
6. Often a very poisonous mushroom such as *Amanita bisporigera* resembles **very** closely an edible mushroom. One such mistake may be fatal.

7. Cooking may reduce the toxin content. Raw toxic mushrooms are probably more dangerous than those that have been cooked.

No book or brief discussion can assure anyone that he or she will or will not be affected by any given mushroom. Each person must respect the potent diversity of mushrooms. The best rules continue to be:

1. Identify mushrooms carefully by using good mushroom books with accurate keys. **Be absolutely certain of the identity of the mushrooms you eat**.

2. Do **NOT** eat *Amanita* species.

3. Eat only a small amount of a mushroom species that you are eating for the first time. Do not eat too much at any one time. Some mushrooms can be eaten in smaller quantities, but in larger quantities they may be deadly.

4. Consider possible allergic reactions; mushrooms may sensitize a person and the next meal will result in serious problems.

5. Consider the **possibility** that mushrooms from lawns that have been sprayed with pesticides may either have been sprayed or possibly have taken up the pesticide from the underground mycelium. We have only recently become aware of this concern. There is no solid evidence that this occurs, but it may be worth considering with the old axiom, "better be safe than sorry."

Mushroom Habitats

The mushroom hunter should be aware of habitats, for fungi grow effectively only where the food they can use is available. The particular species of plants of an area, the association of plants (a prairie, an upland deciduous forest, a flood plain, a grove of planted pines), the soil and other geological factors, rainfall patterns, and the season of the year may influence not only the survival of a particular fungus species but may also affect the production of fruiting bodies by that fungus in any given year.

Mushroom collectors use their knowledge of fruiting seasons and habitats when hunting fungi. Some fungi have a short, rather well defined fruiting time each year; other species have a longer, less definite fruiting time; while some can be sought and found anytime from spring through fall.

The method of obtaining food and nutrients for survival and reproduction delimits the various habitats of fungi. Some fungi, including many mushrooms, are saprobes; i.e., they obtain the energy they need to grow only from the digestion of decaying organic matter. The enzymes that a particular fungus can produce delimit the organic compounds that can be utilized. A given fungus may be able to utilize only a few, or perhaps only one, of the organic compounds available, while another fungus may utilize another compound or group of compounds. Thus, a particular substrate may support the growth of a number of fungi simultaneously or in sequence. As each fungus species accumulates enough food reserves under appropriate conditions, it produces its characteristic spore-producing structures (mushroom, cup, branched coral, puffball).

Other fungi are parasites and get their food from living hosts, plants or animals, at the expense of the host. These parasitic fungi vary widely in the methods of establishing food relationships; they may produce toxic compounds that kill living cells, which are then used for food; they may establish special arrangements with living cells that enable the fungus to obtain food directly from the host cells without killing those cells; or they may live and develop only in particular kinds of cells in the plant, such as the water-conducting cells in the veins of the plant.

Then there are fungi that establish special relationships with the roots of living plants (a mycorrhiza) in which both the host plant and the fungus are mutually benefitted. The fungus grows between or within the root cells and into the surrounding soil; the form of the root may or may not be changed. The fungus utilizes materials in the root cells, and the fungal filaments enhance absorption of mineral nutrients from the soil. These symbiotic mycorrhizal relationships can be so specific that, in some instances, a fungus can live only in association with one species of vascular plant or a species of plant can thrive only in association with a specific fungus. However, in other instances fungi may establish successful mycorrhizal relationships with several, or even a broad range of, vascular plant species.

The following outline summarizes the general habitat and time of fruiting of some of the very common fungi discussed in this book.

  I. On ground in woods
    A. Spring
        1. Morels: *Morchella* spp.
        2. False morels: *Gyromitra* spp., *Verpa* spp.
    B. Summer
        1. Mushrooms: *Amanita* spp., *Amanita vaginata*, *Clitocybe* spp., *Cortinarius* spp., *Entoloma* spp., *Lactarius* spp., *Laccaria laccata*, *Oudemansiella radicata*, *Russula* spp.
        2. Boletes: *Boletus edulis*, *Gyroporus castaneus*, *Strobilomyces floccopus*, *Tylopilus felleus*, *Xerocomus chysenteron*
        3. Saddle fungi: *Helvella crispa*, *Helvella elastica*
        4. Cup fungi: *Peziza succosa*

5. Corals: *Ramaria* spp., *Clavaria* spp.
6. Others: *Cantharellus cibarius, Craterellus cornucupoides, Hydnum repandum, Tremellodendron* spp.
C. Fall
  1. Mushrooms: many of above mentioned summer genera as well as *Tricholoma* spp., *Hygrophorus russula*
  2. Boletes: *Boletus edulis, Strobilomyces floccopus, Tylopilus felleus, Suillus* spp.
  3. Puffballs: *Calvatia gigantea, Calvatia* spp., *Lycoperdon* spp.
II. On ground in woods, but attached to buried plant parts such as sticks
  A. Spring: *Urnula craterium, Sarcoscypha coccinea*
  B. Summer: *Sarcoscypha floccosa, Sarcoscypha occidentalis, Galiella rufa*
  C. Late fall: *Sarcoscypha coccinea*
III. On down wood such as logs, twigs, and branches on the ground in woods
  A. Spring
    1. Mushrooms: *Flammulina velutipes, Pleurotus ostreatus*
    2. Poroid fungi: *Polyporus squamosus, Polyporus arcularis*
  B. Summer
    1. Mushrooms: *Coprinus quadrifidus, Crepidotus* spp., *Marasmius rotula, Mycena leaiana, Mycena* spp., *Phyllatopsis nidulans, Pholiota* spp., *Pleurotus ostreatus, Pluteus cervinus*
    2. Coral fungi: *Clavicorona pyxidata, Ramaria stricta*
    3. Jelly fungi: *Tremella* spp., *Auricularia auricula*
    4. Bird's nest fungi: *Cyathus striatus, Crucibulum laeve*
    5. Cup fungi: *Peziza repanda, Scutellinia scutellata*
    6. Others: *Schizophyllum commune*
    7. Slime molds: *Fuligo* or *Mucilago* spp., *Lycogola epidendrum, Stemonitis* spp., *Physarum* spp.
  C. Fall
    1. Mushrooms: *Pleurotus ostreatus, Pleurotus serotinius, Pholiota* spp., *Pluteus cervinus*
    2. Miscellaneous others: *Hericium coralloides, Hericium erinaceus, Schizophyllum commune, Coriolus versicolor, Coriolus* spp., *Polyporus picipes, Laetiporus sulphureus, Ischnoderma resinosum, Stereum* spp., *Bisporella citrina*
IV. Around or on stumps or where roots of stumps are decomposing in the ground
  A. Summer: *Coprinus micaceus, Coprinus atramentarius, Phallus impudicus, Panus rudus, Pluteus cervinus, Geastrum triplex*
  B. Fall: *Armillariella mellea, Omphalotus illudens, Pluteus cervinus, Laetiporus sulphureus, Grifola frondosa*

V. Standing dead trees
    A. Spring and fall: *Pleurotus ostreatus, Flammulina velutipes, Polyporus* spp., *Fomes* spp.
VI. Living trees
    A. Summer: *Fomes* spp., *Laetiporus sulphureus, Volvariella bombycina*
    B. Fall: *Fomes* spp., *Pleurotus ulmarius*
VII. Pastures and lawns
    A. Summer:
        1. Associated with various dead organic materials in soil: *Marasmius oreades, Agaricus campestris, Chlorophyllum molybdites, Lepiota naucina, Psathyrella candolleana, Coprinus comatus*
        2. Associated with dead or dying tree roots in soil: *Phallus impudicus, Coprinus micaceus, Coprinus atramentarius*
    B. Fall: *Chlorophyllum molybdites, Coprinus comatus, Lycoperdon* spp., *Calvatia* spp., *Scleroderma polyrhizon*
VIII. Parasitic on other fungi
    A. Summer and fall
        1. On mushrooms (probably *Russula* spp. or *Lactarius* spp.): *Hypomyces lactifluorum, Asterophora lycoperdoides*
        2. On boletes: *Sepedonium chrysospermum*
        3. On old fungus fruiting bodies: *Hypomyces rosellus,* particularly on *Tremella* spp.

We have included field characters, habitat notation, spore measurement, and spore print color as factors that determine species; ordinarily these features effectively identify the fungus. Additional microscopic observations of value for the professional mycologist but not included here, are morphology, size, and content of cystidia; morphology of the hyphal cells of the cap; presence of clamp connections; and various other microscopic details. Chemical tests such as those to identify starch in spores and hyphal cells, oil deposits, and color changes in response to specific chemical solutions may be useful. For example, Melzer's solution (see Glossary for the formulation) is a commonly used reagent that permits distinction between amyloid (blue, starch-containing compounds), nonamyloid, and dextrinoid (reddish brown-staining compounds) deposits in spores or hyphae. Other chemical tests may be found in several of the books listed in the General References. These additional tests will be encountered in many of the more technical books and technical articles dealing with mushrooms.

Keys

WHILE keys may look unfamiliar and formidable, their use is learned quite easily. Each numbered couplet in the key (i.e., 1a and b) is a set of alternative descriptions, usually an either/ or choice based on a macroscopic field character. Each choice includes another number that leads to another set of alternatives for consideration. One proceeds through the key choosing alternatives that describe the features found in the fungus in question until finally a name, a page number, or both are given; a general description is given on that page.

The following key will lead the reader to the major groups of fungi included in this guide; where only one species is included, the key will be to that species. In the sections on mushrooms and boletes, there are supplemental keys to the genera in addition to the keys to the species within each family discussed. In each of the subsequent sections there is a key to the genera and species except where there are so few species discussed that a key is not necessary. Many mushrooms are not included in these keys or in this book; thus you may come to dead end. A more serious problem could be the misidentification of a fungus not included here. There is no satisfactory solution to this problem if you use only this guide.

KEY TO MAJOR GROUPS OF FUNGI INCLUDED IN THIS GUIDE

A. Fruiting body producing asci or basidia that actively release spores into the air; a spore print can be made.

    1a. Fruiting body fleshy; cap with thin gills (plates of tissue) on lower surface; usually with stalk—Agaricales (Mushrooms) 2
For fleshy fruiting bodies with broad ridges on lower surface . . .
. . . . .Cantharelloid Aphyllophorales, p. 171

For tough whitish fruiting bodies that dry but revive with moisture and have longitudinally split "gills" on lower surface . . . . .Schizophylloid Aphyllophorales (*Schizophyllum commune*), p. 208

  b. Fruiting bodies of various textures; lacking thin gills 14

2a. Spore print white to cream or green 3
  b. Spore print other than white, cream, or green 7

3a. Gills entirely free from stalk 4
  b. Gills partly or entirely attached to stalk 5

4a. Volva present at base of stalk . . . . .Amanitaceae, p. 47
  b. Volva not present at base of stalk . . . . .Lepiotaceae, p. 86

5a. Gills thick and fleshy, waxy when rubbed . . . . .Hygrophoraceae, p. 79
     See also. . . . . .Tricholomataceae (*Laccaria*), p. 133
  b. Gills thin or thick but never waxy 6

6a. Cap flesh brittle, with or without latex . . . . . .Russulaceae, p. 98
  b. Cap flesh firm, soft, tough (not brittle), no latex. . . . . .Tricholomataceae, p. 121

7a. Spore print salmon to pink 8
  b. Spore print neither pink nor salmon. .9

8a. Gills free; volva present or not; growing on wood or in soil . . . . .Volvariaceae, p. 150
  b. Gills attached; volva absent; growing in soil . . . . .Rhodophyllaceae, p. 93

9a. Spore print smoky gray to black 10
  b. Spore print other than gray or black. 11

10a. Gills thick, somewhat fleshy, extending along stalk (decurrent); on ground near or under conifers. . . . . .Gomphidiaceae, p. 78
  b. Gills thin, often breaking down into inky mass, not decurrent; on wood in soil or on rotting wood . . . . .Coprinaceae, p. 62

11a. Spore print purple-brown to chocolate brown 12
  b. Spore print other than purple-brown or chocolate brown 13

12a. Spores usually purple-brown; gills white to tan when young, attached to stalk. . . . . .Strophariaceae, p. 115

b. Spores usually chocolate brown; gills gray, sometimes white or pink when young, free from stalk . . . . .Agaricaceae, p. 41

13a. Spores bright yellow to clay-brown; gills usually free. . . . .
. . . . .Bolbitiaceae, p. 59
b. Spores rust-brown to cinnamon brown; gills attached . . . .
. . . . .Cortinariaceae, p. 73

14a. Fruiting body caplike with stalk or bracketlike; lower surface with a layer of tubes opening as pores or with a layer of smooth or jagged spines 15
b. Fruiting body variously shaped; lower surface without tubes or spines 17

15a. Fruiting body with conical or jagged spines on lower surface . . .
. . . . .Spiny Aphyllophorales, p. 212
b. Fruiting body with tube layer on lower surface 16

16a. Fruiting body a fleshy cap; usually with central stalk; on ground associated with live trees, usually mycorrhizal. . . . .
. . . . .Agaricales (Boletes), p. 155
b. Fruiting body seldom fleshy, usually tough, corky, or woody, often bracketlike and broadly attached; occasionally with stalk; directly attached to living or dead trees . . . .
. . . . .Poroid Aphyllophorales, p. 184

17a. Visible fungus structures developed on diseased living plants. . . .
. . . . .Plant parasitic fungi, p. 289
b. Fruiting bodies of various sizes and textures; not directly associated with living plants, usually on ground or with dead plant parts 18

18a. Fruiting bodies with a stalk and a differently shaped upper portion 19
b. Fruiting bodies lacking a definitely different stalk 28

19a. Fruiting bodies fleshy 20
b. Fruiting bodies tough, fleshy to leathery, to hard 25

20a. Fruiting bodies much branched, single, or in small clusters; on ground in woods or on rotting logs . . . .
. . . . .Clavaroid Aphyllophorales (Branched Corals), p. 177
b. Fruiting bodies not branched, single, or in groups; on the ground or parasitic 21

21a. Fruiting bodies with cap and stalk (mushroomlike); lower surfaces with broad ridges extending along stalk, or smooth.........22
  b. Fruiting bodies with a differently shaped upper area........23

22a. Fruiting bodies firm, often irregular in shape, bright orange becoming water soaked and reddish purple with age; covered with slightly raised, rounded areas with central water-soaked dots— Mushrooms parasitized by..............................
  ....*Hypomyces lactifluorum*, p. 287
  b. Fruiting bodies not undergoing such color change with age; lower surface smooth or with broad ridges extending along stalk.....
  ....Cantharelloid Aphyllophorales
  (Chantarelles and associates), p. 171

23a. Fruiting bodies with clublike upper area.................24
  b. Fruiting bodies with pitted spongelike upper portion, or with convoluted apical region, or saddle-shaped area................
  ....Ascomycetes, p. 251

24a. Fruiting body growing from a parasitized insect pupa or larva buried in soil or in a rotting log........*Cordyceps* spp., p. 282
  b. Fruiting body not associated with a juvenile insect stage.......
  ....Clavaroid Aphyllophorales (Single Corals), p. 177

25a. Fruiting bodies hard, club shaped, single, or with few branches; solid white central region covered by black rind; associated with wood........*Xylaria polymorpha* (Dead Man's Fingers), p. 288
  b. Fruiting bodies leathery to tough fleshy; many branches, on ground...26

26a. Fruiting body tough fleshy; branches hollow, rounded........
  ....*Tremella reticulata*, p. 249
  b. Fruiting bodies leathery; branches solid, flattened..........27

27a. Fruiting bodies light colored, sometimes overgrown by green algae; in groups on ground in woods......................
  ....*Tremellodendron* spp., p. 250
  b. Fruiting bodies purple-brown; tips of young branches whitish to gray; on ground...................*Thelephora* spp., p. 250

28a. Fruiting bodies thin leathery brackets, broadly attached, often in overlapping clusters; lower surface smooth; on wood.........
  ....Smooth Aphyllophorales (*Stereum* spp.), p. 210
  b. Fruiting bodies not as above.........................29

29a. Fruiting bodies black, hard and carbonous, rounded to flattened; on down wood or occasionally on standing trees . . . .
. . . .*Hypoxylon* spp., p. 285
Or . . . .*Daldinia concentrica*, p. 285
  b. Fruiting bodies at least somewhat fleshy, not carbonous 30

30a. Fruiting bodies flat, or indefinite size, light or bright yellow; attached firmly on down wood. . . . . .*Hypocrea sulphurea*, p. 287
  b. Fruiting bodies upright, lobed, or cup- or saucerlike; on wood or on the ground. 31

31a. Fruiting bodies firm gelatinous or hard and brittle when dry, becoming firm gelatinous when remoistened . . .Jelly Fungi, p. 243
  b. Fruiting bodies somewhat fleshy, cup- or saucer-shaped . . . .
. . . . .Cup Fungi, p. 251

B. Fruiting bodies producing dry powdery spores in covered, enclosed masses or producing slimy spore masses.

  1a. Fruiting bodies may be fleshy when young, with dry powdery spore masses when mature. 2
  b. Fruiting bodies fleshy with slimy, stinking spore masses over apical portion . . . . .Gasteromycetes (Stinkhorns), p. 217

  2a. Fruiting bodies tough, cylindrical to goblet-shaped, nestlike, containing a number of lens-shaped "eggs"; spores inside the eggs and released only when tough outer covering decays or breaks . . . . .
. . . . .Gasteromycetes (Bird's Nest Fungi), p. 217
  b. Fruiting bodies containing dry powdery spore mass readily exposed when outer covering breaks 3

  3a. Fruiting bodies small (about 1 cm in diameter) and few or very small (1–3 mm in diameter) and many in a given area; cover usually breaking, soon releasing spores . . . .
. . . . .Myxomycetes (Slime Molds), p. 295
  b. Fruiting bodies larger (more than 1 cm in diameter); single or scattered . . . .
. . .Gasteromycetes, (Puffballs, Earthstars, Hard Puffballs), p. 217

Mushrooms

THE mushrooms are a group of fleshy fungi that have gills, thin plates of tissue lined on both sides with basidia, on the lower surface of the cap. Details of their basic structure and development are presented in the Introduction. While common usage of the term mushroom may include other fleshy fungus fruiting bodies, such as those of morels, tooth fungi, and sometimes even puffballs, we are using the term in the more limited, technical interpretation.

Another source of confusion among terms used for fungus fruiting bodies is the term toadstool. Originally intended to indicate those mushrooms that are inedible or toxic or poisonous, the term cannot be used with precision. There is no basis, either technically or on field information, to describe such a group as distinct from mushrooms in general.

KEY TO THE GENERA OF MUSHROOMS INCLUDED IN THIS GUIDE

1a. Fruiting body attached directly to another mushroom species, parasitic; cap becoming a powdery mass of spores . . . .
. . . .*Asterophora*, p. 127
  b. Fruiting body not attached directly to another mushroom species, not parasitic 2

2a. Fruiting body shelflike, broadly laterally attached or attached by short lateral stalk; on wood. 3
  b. Fruiting body with stalk attached centrally, off-center, or laterally . .8

3a. Spore deposit white to pale yellowish or pale lilac 4
  b. Spore deposit pink or some shade of brown 6

4a. Gill edges serrate (sawtoothlike) . . . .*Lentinellus*, p. 146
  b. Gill edges smooth 5

5a. Fruiting body tough, reviving when moistened after drying . . . .
. . . .*Panus*, p. 143
b. Fruiting body somewhat tough to soft, not reviving when moistened . . . .*Pleurotus*, p. 145; *Panellus*, p. 146

6a. Spore deposit yellow-brown to brown . . . .*Crepidotus*, p. 74
b. Spore deposit pink 7

7a. Fruiting body sessile; gills bright orange . . . .*Phyllotopsis*, p. 143
b. Fruiting body usually with short stalk; upper surface of cap wrinkled, pinkish . . . .*Rhodotus*, p. 146

8a. Gills dissolve as fruiting body matures; spores black . . . .
. . . .*Coprinus*, p. 65
b. Gills remain as definite structures as fruiting body matures; spores variously colored 9

9a. Fruiting body with universal veil that develops into a volva at maturity and/or a partial veil, often becoming a ring or annulus at maturity 10
b. Fruiting body lacking these structures 21

10a. Volva present, gills free 11
b. Volva lacking, gills free or attached 12

11a. Spore deposit white; ring present . . . .*Amanita*, p. 47
b. Spore deposit pink to dingy pink; ring lacking . . .*Volvariella*, p. 153

12a. Annulus conspicuous, persistent at maturity 13
b. Annulus present or lacking, often not conspicuous at maturity . . .16

13a. Spore deposit light green . . . .*Chlorophyllum*, p. 87
b. Spore deposit not green 14

14a. Spore deposit white . . . .*Lepiota*, p. 89; *Leucoagaricus*, p. 92
b. Spore deposit some shade of brown 15

15a. Spore deposit purple-brown to blackish brown . . . . .*Agaricus*, p. 43
b. Spore deposit rusty to yellowish . . . .*Rozites*, p. 77

16a. Spore deposit white . . . .*Armillariella*, p. 125
b. Spore deposit colored 17

17a. Spore deposit yellowish clay to rusty brown 18
b. Spore deposit blackish, violaceous brown to chocolate brown . . .19

18a. Partial veil cobwebby on young fruiting bodies, remnants visible for a while on stalk; typically on ground in woods . . . . *Cortinarius,* p. 74
  b. Partial veil remains as annulus, usually well developed and/or stalk scaly; cap scaly or smooth . . . . *Pholiota,* p. 117

19a. Gills decurrent; on ground under or near conifers. . . . .
  . . . . *Gomphidius,* p. 78
  b. Gills attached but not decurrent. 20

20a. Annulus persistent . . . .*Stropharia,* p. 119
  b. Annulus lacking . . . .*Psathyrella,* p. 71

21a. Gills thick, waxy, subdistant. 22
  b. Gills thin, not waxy 23

22a. Gills flesh color to purple; spore deposit white or pale lilac; spores spiny . . . . *Laccaria,* p. 133
  b. Gills white, occasionally yellowish or orange; spore deposit white; spores smooth . . . . *Hygrophorus,* p. 80

23a. Cap and gills brittle, crumble easily; on ground in woods or near trees 24
  b. Cap and gills firm, do not crumble; on ground, wood, or litter. . .25

24a. Milky juice present, may be difficult to obtain from old fruiting bodies. . . . .*Lactarius,* p. 101
  b. Milky juice always absent . . . . *Russula,* p. 109

25a. Fruiting bodies on, or associated with, wood. 26
  b. Fruiting bodies on ground on various kinds of organic debris. . . .32

26a. Spore deposit white 27
  b. Spore deposit colored 31

27a. Gill edges serrate (sawtoothlike) . . . .*Lentinus,* p. 135
  b. Gill edges smooth 28

28a. Cap margin straight; cap usually bell shaped . . . . . .*Mycena,* p. 139
  b. Cap margin incurved when young, expanding with age 29

29a. Gills decurrent . . . .*Omphalotus,* p. 141
  b. Gills attached, not decurrent 30

30a. Stalk velvety, dark brown at maturity; typically in clusters. . . . .
  . . . . *Flammulina,* p. 133

Agaricaceae

Mushrooms in this family are usually fleshy with rounded to flattened caps, which range in color from white to brown or gray-brown. Cap surface is smooth to scaly and the flesh normally white to pink, sometimes changing color when cut or bruised.

Gills are free, white to gray or pink to dull chocolate brown to purple-brown.

Stalks are relatively thick, always with a prominent annulus, either single or double. Stalk usually smooth above the annulus, often hairy to scaly below.

Spores are purplish brown to dark brown, elliptical, smooth, with an apical pore; spore print dark chocolate brown to purplish brown.

The species occur in a number of habitats including lawns and pastures, humus and litter in hardwood and conifer forests, and manured areas. They often resemble the white Amanitas in early development, thus it is important to examine the entire mushroom.

Many of the species of this group are prized edible mushrooms, but there have been reports of gastrointestinal upsets, particularly with *Agaricus silvaticus*. These may be individual reactions and not a general toxicity, but it is wise to eat sparingly of any species until you know your individual reaction.

KEY TO THE SPECIES OF *AGARICUS* IN THIS GUIDE

(Be aware of misidentification. Do NOT eat any specimen not positively identified.)

1a. Caps white, at least in young specimens 2
  b. Caps having some color. 4

2a. Partial veil single 3
  b. Partial veil double. . . . .*A. silvicola*

3a. Flesh turning yellowish when cut or bruised . . . .*A. arvensis*
  b. Flesh unchanged in color when cut or bruised. . . . .*A. campestris*

4a. Cap gray, covered with patches of blackish brown, dense, hairy scales. . . . .*A. placomyces*
  b. Cap reddish brown, covered with pinkish brown, hairy scales . . . . . .
  . . . .*A. silvaticus*

19 *Agaricus arvensis*

20 *Agaricus campestris*

Agaricus arvensis Fr. (19)   Edible, choice
"HORSE MUSHROOM"

Cap 7–15 cm broad, convex, flattening at maturity; white to cream with center yellowish brown; dry and smooth except center with small scales. Flesh thick; white slowly yellowing when cut.

Gills free, crowded; initially white to gray-pink to deep chocolate brown at maturity.

Stalk 6–12 cm long, 1–2 cm thick, sometimes with a bulb at base; white bruising yellow; smooth, dry above the annulus, cottony scales below annulus.

Partial veil two layered, membranous; white; outer layer breaking into toothlike pattern; annulus skirtlike.

Spores 7–9 × 4–6 μm, smooth, elliptical; spore print dark purplish brown.

Scattered in pastures and fields or lawns after heavy rains, S., F.

This is a common species in some areas. Sometimes confused with *A. campestris*, *A. arvensis* typically stains yellow upon bruising or cutting of cap or stalk. Spores are larger, more nearly elliptical rather than ovoid as in *A. campestris*. Another similar species, *A. silvicola*, found in woodland habitats, has similar general morphology, but smaller spores (5–6.5 × 3–4 μm).

Agaricus campestris L.: Fr. (20)   Edible, choice
"MEADOW MUSHROOM"

Cap 3–10 cm broad, convex to nearly flat at maturity; white becoming light brown or darker; smooth to fibrillose. Flesh firm; white or brownish, unchanging or slowly yellowing when cut.

Gills free, crowded, narrow; pink maturing to purple-brown.

Stalk 3–6 cm long, 1–2.5 cm thick, equal; white and smooth above the annulus, white and hairy below, cap color at maturity.

Annulus a single layer, thin, sometimes disappearing.

Spores 6–7.5 × 4–5 μm, ovoid to elliptical; spore print dark chocolate brown.

Scattered to abundant in lawns, meadows, and pastures following rains and cool weather, Sp., S., F.

This is a common, delicious mushroom and is closely related to the cultivar *A. brunnescens*, which is grown commercially. The characteristic bright pink, free gills and the brown spore print clearly distinguish this mushroom from the white-spored Amanitas. A similar species, also edible, is *A. bitorquis*, also known as *A. rodmani*, which differs from *A. campestris* by having a hairy, double-layered annulus, spores 4–6.5 × 4–5 μm, and a very short stalk. *A. bitorquis* is usually found in compacted soil near paths, sidewalks, and streets.

21   *Agaricus placomyces*

Agaricus placomyces Pk.   **(21)**   Edibility questionable

Cap 4–10 cm broad, oval to rounded, maturing to almost flat; white with gray to brown hairy scales; margin incurved when young, becoming straight at maturity. Flesh white to pale pink to dull pink bruising yellow and finally reddish brown.

Gills free, crowded; gray-pink maturing to chocolate brown.

Stalk 3–9 cm long, 0.5–2 cm thick, slightly enlarged toward base; white becoming dingy brown with age.

Annulus membranous; cottony white with yellow to brown; soft, flat patches beneath, persistent.

Spores 5–6 × 3.5–5 μm, ovoid to elliptical, smooth; spore print chocolate brown.

Scattered to loosely clustered under hardwoods and in leaf piles, late S. or F.

This species seems to be edible for some people but poisonous for others. It probably should be **avoided.**

Agaricus silvaticus Fr.   Not recommended

Cap 4–12 cm, ovate to flattened; dingy pinkish brown to dark brown; fibrils grouped into clusters. Flesh firm; white unchanging or slowly bruising reddish brown.

Gills free, crowded; purplish gray to dull pink becoming purple-brown.

Stalk 6–11 × 1–2 cm, somewhat silky; base subbulbous.

Annulus with floccose scales, patches or fibrils on lower surface, persistent.

Spores 5–6 × 3.0–3.4 μm, ovoid, smooth; spore print chocolate brown.

Scattered under hardwoods, late S., F.

A. silvaticus and *A. silvicola* occur under hardwoods in similar habitats. However, *A. silvicola* lacks scales on the cap, and the cap flesh bruises yellow rather than reddish brown as does *A. silvaticus.*

22 *Agaricus silvicola*

Agaricus silvicola (Vitt.) Sacc. **(22)**   Not recommended

Cap 6–15 cm broad, convex becoming flat at maturity; white to creamy white; somewhat silky fibrillose. Flesh firm, moderately thick, brittle; white staining yellow when bruised; odor mild.

Gills free, crowded, narrow to moderately broad; white at first then pink and finally chocolate brown.

Stalk 7–15 cm long, 1–2 cm thick, sometimes tapering slightly upward; white staining yellow when bruised; base flattened or with a bulb.

Annulus large, double, smooth above with lower surface breaking into yellowish patches that may disappear.

Spores 5–6.5 × 3–4.5 μm, elliptical, smooth; spore print purplish brown to chocolate brown.

Scattered or in clusters, under hardwoods, S., F.

A. silvicola is quite variable, both in size and in stalk base. The annulus is large and conspicuous. Reports on edibility vary also, therefore we do **not** recommend it.

Amanitaceae

All mushrooms in this group develop from an oval to round structure (button) covered by a universal veil, which breaks when the mushroom stalk elongates. Caps are rounded to conic or flattened at maturity, and may be smooth, viscid, or warty from universal veil remnants. Cap margins are often inrolled at first and entire or striate. Flesh is white to cream, either unchanging or bruising to brown or red in some instances.

Gills are free, white or nearly so, and close to crowded.

The stalk is even, smooth to roughened toward the base by universal veil remnants, and may be enlarged to a bulbous base enclosed by the volva. Annulus is usually membranous if present, or absent. Volva may be cottony or membranous in *Amanita* or gelatinous in *Limacella*. *Limacella* is not common in the Midwest and is not included in this guide.

Spores are oval to round, smooth and thin walled, amyloid or nonamyloid in Melzer's reagent, and the spore prints are white to cream.

A mushroom can be recognized in the field as a species of *Amanita* if it has white spores, free gills, and a volva.

This family, which includes the **most** poisonous mushrooms, accounts for most of the deaths due to mushroom poisonings in the United States. They are, however, attractive and of interest because of their mycorrhizal associations. They are also of interest to both amateur and professional mycologists if for no other reason than to avoid them when collecting edible mushrooms. Finally, there is a fascination with the deadly members of our environment, such as a cobra, a white shark, or the "destroying angel" mushroom.

Certain species of *Amanita* are sometimes labeled as edible in mushroom books. In spite of this, we urge you **NOT** to eat any species of *Amanita* **NOR** any mushroom similar in appearance.

KEY TO THE SPECIES OF *AMANITA* IN THIS GUIDE

(Be aware of possible misidentification. Do NOT eat any specimen not positively identified.)

1a. Annulus absent; cap white to gray to gray-brown, or light orange, striate . . . .*A. vaginata* group
 b. Annulus present; cap white or colored 2

2a. Fruiting body entirely white; volva a loose saclike sheath . . . .
 . . . .*A. bisporigera* group
 b. Cap with some color; volva variously developed 3

3a. Cap whitish with some yellowish olive, at least at center . . . .
. . . .*A. cothurnata*
 b. Cap a definite color other than described above 4

4a. Cap brown, red-brown, gray-brown, or yellow-brown. 5
 b. Cap other than shades of brown 7

5a. Cap flesh and stalk not staining when bruised or cut . . . .*A. velatipes*
 b. Cap flesh and stalk staining reddish or brownish. 6

6a. Stalk with an abrupt expanded bulb with a depressed margin; margin
often split longitudinally. . . . .*A. brunnescens*
 b. Stalk with clavate to ovoid base, not splitting . . . .*A. rubescens*

7a. Cap orange or red, at least in central area .8
 b. Cap yellow, yellow-green, or yellowish buff 9

8a. Volva breaking in pieces around base of stalk; cap margin even or
faintly striate . . . .*A. flavoconia*
 b. Volva persistent; cap margin striate . . . .*A. muscaria*

9a. Cap yellowish green; margin not striate. . . . .*A. citrina*
 b. Cap pale yellow to yellowish buff; margin striate. . . . . . .*A. gemmata*

Amanita bisporigera Atk.   (23)   Deadly poisonous
"DESTROYING ANGEL"

Cap 4–12 cm broad, rounded or conic to flat at maturity, smooth; pure white; viscid or sticky when wet; margin at first inrolled, lacking striations. Flesh firm; white, outer cap layer yellow when KOH is applied.

Gills free; white.

Stalk 10–20 cm long, 1–2.5 cm thick, tapering from enlarged base to apex; white, barely fibrous above annulus, distinctly hairy below.

Annulus membranous, smooth; white; irregular, a delicate skirtlike ring that may **easily detach** from the stalk.

Volva membranous; white; a distinct, persistent cup surrounding but separate from the inflated base.

Spores only 2 per basidium, 7–10 μm, globose, amyloid in Melzer's reagent, thin walled; spore print white.

Scattered under hardwoods, S. or F.

The description of *A. virosa* is the same except there are 4 spores per basidium, and the spores are 8.5–11 μm or 9–11 × 7–9 μm globose to subglobose. The cap flesh turns yellow when KOH is applied. The description of *A. verna* is also the same as *A. bisporigera* except the spores are 9–12 × 6.5–8 μm, and the cap flesh is not yellow when KOH is applied.

Amanita bisporigera, *A. virosa,* and *A. verna* can be distinguished from one another only by spore characteristics. All three species are equally **poisonous** and are to be **avoided** by anyone seeking edible fungi. *A. virosa* has been reported to be responsible for most of the mushroom deaths in the United States. *A. bisporigera* is the most common of the three species in Iowa.

23   *Amanita bisporigera*

24 *Amanita brunnescens*

Amanita brunnescens Atk. (24)   Poisonous

Cap 2.5–12 cm broad, rounded to flat at maturity; gray-brown to dark brown with white patches when young; viscid; margin faintly striate. Flesh thin; white bruising reddish brown.

Gills free, crowded; white.

Stalk 5–13 cm long, 1–2 cm thick, expanded to a large bulbous base, which has longitudinal splits or crevices; white staining reddish brown when bruised; smooth above ring, hairy below.

Annulus membranous; white to pallid; may collapse around stalk.

Volva remains as patches on the cap and on the basal bulb sometimes forming an inconspicuous collar around top of the basal bulb.

Spores 7–10 μm round, thin walled; spore print white.

Scattered under hardwoods and mixed woods, S., F.

The lengthwise splits in the basal bulb and the viscid brown cap are good field characteristics for this species.

Amanita citrina (Schaeff.) S. F. Gray   Poisonous

Cap 5–12 cm broad, rounded to broadly knobbed; greenish yellow to light olive buff with irregular, flattened white to creamy patches of universal veil; viscid; margin light. Flesh soft; white.

Gills free but reaching stalk, crowded to close; white.

Stalk 8–10 cm long, 1–1.5 cm thick enlarging to the nearly round basal bulb, often one to several longitudinal splits in the bulb; white; smooth above and cottony below the annulus.

Annulus membranous, fragile; cream to buff; persistent, often collapsing around the stalk.

Volva flattened against basal bulb, evident at the rim of the bulb as a distinct collar.

Spores 7–9 μm, nearly round, amyloid; spore print white.

Single to scattered under conifers, mixed woods, or hardwoods, S., F.

The patches of universal veil on the cap often wash off in rainy weather, thus this feature is not a reliable characteristic for identification.

25 *Amanita cothurnata*

26 *Amanita flavoconia*

Amanita cothurnata Atk. (25) Poisonous

Cap 4–8 cm broad, rounded to flattened; white to tan or yellowish olive, darker in center; viscid to dry with numerous white, woolly, moderately persistent warts or patches; margin faintly to conspicuously striate. Flesh white.

Gills free, close; bright white.

Stalk 5–10 cm long, 1–2 cm thick, nearly equal to slightly tapered upward; white; smooth above to somewhat floccose below the annulus; base bulbous, enlarged, hollow.

Annulus floccose, membranous; white; usually attached near to slightly above the middle of the stalk, sometimes slightly flaring.

Volva usually close to the stalk base, forming a distinct narrow collar.

Spores 8–10 × 6–8 μm, broadly elliptical to nearly round, nonamyloid; spore print white.

Single to scattered in hardwoods, S., F.

The white to creamy cap with a darker center, the striate margin, and the close-fitting volva with a collar on the top of the basal bulb are characteristic. This species is one member of a group of *Amanita* species, including *A. muscaria*, with warts and patches, remnants of the universal veil on the cap; a striate cap margin; and often a distinct collar at the top of the basal bulb. *A. velatipes* may be quite similar in appearance.

Amanita flavoconia Atk. (26) Probably poisonous

Cap 3–9 cm broad, rounded to nearly flat; red-orange to yellow-orange; large, cottony, yellow warts; viscid; margin even to faintly striate. Flesh thin, firm, white to cream to nearly buff near the cap surface.

Gills free, close; white with finely hairy edges.

Stalk 5–10 cm long, 0.5–1.5 cm thick, enlarging to an almost oval basal bulb; yellow, powdery above annulus, white to pale yellow below annulus.

Annulus membranous; white to cream; persistent.

Volva breaking up around the bulb and lower part of stalk, usually fragmenting in the soil when mature; bright yellow-gold.

Spores 7–9 × 4.5–6 μm, elliptical, amyloid; spore print white.

Single to scattered under hardwoods and mixed forests, S., F.

Amanita flavorubescens is similar to *A. flavoconia* with a yellow to yellow-brown, viscid cap with yellow warts; however, the lower stem and bulb slowly become reddish. *A. flavoconia* is also easily confused with *A. frostiana*. The globose to subglobose, 7–10 μm, nonamyloid spores of *A. frostiana* are a differential characteristic.

27 *Amanita gemmata*

28 *Amanita muscaria*

Amanita gemmata (Fr.) Gillet **(27)**   Poisonous

Cap 2-7 cm broad, rounded to flattened at maturity; pale yellow with darker center and patches of whitish, irregularly shaped, membranous universal veil remnants; margin distinctly striate. Flesh soft, thin; white to cream.

Gills free, close; creamy white.

Stalk 6–14 cm long, 0.5–2 cm thick, enlarged near basal bulb; white; nearly smooth above to somewhat floccose toward the base.

Annulus membranous, fragile; white; sometimes remaining attached to the margin of the cap.

Volva attached to the basal bulb; white to cream; margin free at first as a collar at the top of the bulb but later breaking down leaving a torn margin.

Spores 8.5–11 × 6.5–9 μm, broadly elliptical, nonamyloid; spore print white.

Single or grouped on the soil in hardwoods or mixed woods, S., F.

Amanita muscaria (Fr.) S. F. Gray **(28)**   Poisonous
"FLY AGARIC"

Cap 7–20 cm broad, round aging to flat with lightly striate edges; ranging from dark red to red-orange to yellow covered with white scaly patches of universal veil remnants; surface viscid when fresh. Flesh white bruising brownish.

Gills free, crowded; white.

Stalk 5–12 cm long, 1.5–3 cm thick enlarging toward the bulbous base; white; bare above ring, hairy or scaly below ring.

Annulus up to 4–5 cm long, membranous, often detaching from the stalk.

Volva remnants appear as irregular concentric rings above the base of the stalk.

Spores 8–12 × 6.5–8 μm, elliptical, thin walled, nonamyloid; spore print white.

Scattered to abundant under conifers and hardwoods, S., F.

This species shows great variation in color as well as in the amounts of toxins present. There are reports of hallucinogenic effects as well as poisonings from many areas. This mushroom should **NOT** be eaten under any circumstances.

29 *Amanita rubescens*

30 *Amanita vaginata*

Amanita rubescens (Pers.: Fr.) S. F. Gray (29)

Not recommended

"BLUSHER"

Cap 5–12 cm broad, rounded or slightly umbonate; reddish brown with shades of gray or tan with irregular, whitish, greenish gray to pinkish warts; viscid to slightly sticky; margin faintly to lightly striate. Flesh initially white staining reddish to red-brown upon bruising or cutting.

Gills close to crowded, free to nearly free; whitish staining quickly to pinkish or reddish brown when bruised or cut.

Stalk 6–20 cm long, 0.8–2 cm thick widening to a swollen basal bulb; lightly flushed with pink above the annulus, commonly stained reddish to red-brown below the annulus.

Annulus membranous; white to pinkish staining reddish; fragile, often tearing irregularly.

Volva breaking into reddish pieces, some adhering to the swollen base, others left in the surrounding soil.

Spores 8–10 × 5–6 μm, elliptical, mostly smooth, strongly amyloid; spore print white.

Singly or in groups under hardwoods, particularly oaks, S., F.

The reddish stains that are the basis for the common name "blusher" and the irregular pinkish gray or greenish gray warts turning reddish on the rounded cap make this a distinctive mushroom. When summer moisture is plentiful this is one of our most striking species. Some daring individuals eat this species, but because of the difficulty in distinguishing between this and some related species that are poisonous, we repeat our **caution** about not eating Amanitas.

Amanita vaginata (Bull.: Fr.) Quél. (30)

Nonpoisonous, not recommended

"GRISETTE"

Cap 3–8 cm broad, conical becoming flat; smoky gray to gray-brown, sometimes with large patches of the universal veil; viscid when fresh; margin conspicuously striate. Flesh thin; white unchanging when bruised.

Gills free, crowded when young; white.

Stalk 8–22 cm long, 1–2 cm thick, slender enlarging to a bulbous base; white; smooth or with sparse, flattened hairs.

Annulus absent.

Volva membranous, saclike but not attached to stalk; white; often buried well below soil line.

Spores 8–12 μm, globose, thin walled, nonamyloid; spore print white. Single to numerous under hardwoods, S., F.

Also known as *Amanitopsis vaginata,* this species is considered nonpoisonous but again we do **NOT** recommend eating. It is easily mistaken for one of the poisonous Amanitas that have lost their universal veil remnants or annulus.

The gray form is the most common in Iowa hardwood forests. It is one of the most common mushrooms, occurring even in years of limited moisture when many mushrooms do not fruit.

Two other varieties of *A. vaginata* are often recognized: *A. vaginata* var. *fulva,* an orange-capped variant, and *A. vaginata* var. *alba,* with a white cap. These have also been recognized as separate species.

Amanita velatipes Atk. **(31)**   Probably poisonous

Cap 6–15 cm broad, rounded to flattened at maturity; creamy yellow darkening to brownish at the center with numerous dingy white, wooly warts; margin striate. Flesh moderately thick at center to thin at margins; white, faintly yellow below the cap cuticle.

Gills broad, free, crowded with many shortened gills of various lengths; creamy white.

Stalk 12–20 cm long, 1–2.5 cm thick, nearly equal tapering slightly upward, stuffed or becoming hollow; creamy white; basal bulb club shaped.

Annulus membranous, large; creamy white; attached at center of stalk and often collapsed around the stalk.

Volva thick, membranous; dingy white to cream; tightly surrounding basal bulb with a somewhat free, rolled margin leaving one or two rings of white floccose tissue on the stalk above the bulb.

Spores 8–10 × 6–7 μm, elliptical, nonamyloid; spore print white.

Scattered to grouped on the soil under hardwoods and in forest openings, S., F.

This mushroom is often observed in close association with the bolete *Tylopilus felleus.* Thus, in our area the two are either mycorrhizal species with the same plant associate or their habitat requirements are nearly identical. *A. cothurnata* may be quite similar in appearance.

31 *Amanita velatipes*

Bolbitiaceae

This family is characterized by bell-shaped to rounded fragile caps somewhat reminiscent of Coprinaceae. Caps range in color from white to pale yellow to tan to darker brown, are moist or viscid to dry, and some have prominent striations at the margin; cap surface layer (cuticle) is cellular. Flesh is thin and white to pallid.

Gills are free to attached and light colored maturing to darker brown.

Stalks are typically long, thin, and slightly enlarged at the base; striate; cap color or lighter; and with or lacking a partial veil.

Spores are yellow-brown to cinnamon or reddish brown, smooth or variously ornamented with an apical pore; spore prints are yellow-brown to darker reddish brown.

Most species are not of interest for edibility because of small size and thin flesh.

32 *Conocybe lactea*

33 *Conocybe tenera*

Conocybe lactea (Lge.) Metrod **(32)**

Cap 1–2.5 cm broad, 1–1.5 cm high, conic; creamy white throughout or slightly darker at top; smooth; margin flaring with age. Flesh very thin.

Gills very narrow, whitish becoming rusty brown.

Stalk 4–8 cm long, 0.1–0.2 thick, equal, very fragile; whitish.

Spores 12–16 × 7–9 μm, truncate; spore print rusty brown.

In groups or scattered in lawns or grassy areas, very common in rainy weather, S., F.

These delicate, fragile mushrooms may be present in large numbers in a lawn on a dewy summer morning and completely wilted and dried by noon.

Conocybe tenera (Schaeff.: Fr.) Kühner **(33)**

Edibility unknown

Cap 1–2 cm broad, 1–2 cm high, conic to bell shaped; reddish to yellow-brown becoming lighter with age; dry, striate nearly to center. Flesh thin, water-soaked; brown.

Gills nearly free, subdistant, narrow; clay-brown to cinnamon brown.

Stalk 4–8.5 cm long, 0.1–0.2 cm thick, straight, equal; cap color; finely striate.

Spores 7–10 × 5–7 μm, elliptical, thick walled, smooth with apical pore; spore print red-brown.

Scattered to abundant on lawns or open grassy areas, Sp., S.

This species is also known as *Galera tenera,* although there is much confusion about this name and the mushroom it signifies. There is a group of small, somewhat fragile mushrooms with conic to bell-shaped caps on long, thin stalks that are common in lawns and grassy areas.

Another species, *Conocybe lactea,* differs in its white to light tan cap. Both species are common in lawns.

Coprinaceae

This family is characterized by thin, typically conical caps with nearly free to adnate (attached) to adnexed (notched) gills and a brown to black spore print. In the genus *Coprinus* the gills are converted to an inky mass by autodigestion, a condition that has led to the name "inky caps" for many species.

The caps range from dry to viscid, some with scales; margins are striate to wrinkled or split. Flesh is initially white or pallid.

Gills are crowded to distant, dissolve or remain firm, and are initially grayish becoming darker. In *Coprinus* they deliquesce to form an inky black mass.

Stalks are relatively thin, smooth to hairy or scaly, and are with or without an annulus.

Spores are usually dark purple-brown to black, elliptical, smooth, with an apical pore; the spore print is dark brown to black.

The Coprinaceae are saprobes on dung, wood debris, or humus in lawns. While several species of *Coprinus* are edible, some have been linked with gastrointestinal upsets. Some species induce other symptoms when they have been eaten with alcoholic beverages. All species of *Paneolus* should be **avoided** because some are reported to contain poisonous or hallucinogenic compounds.

KEY TO THE GENERA AND SPECIES IN THIS GUIDE

(Be aware of possible misidentification. Do NOT eat any specimen not positively identified.)

1a. Gills breaking down into an inky black mass. 2
  b. Gills remaining firm, not inky or black becoming darker as spores mature 5

2a. Cap long, narrowly cylindric conical, white, covered with recurved brownish scales . . . .*Coprinus comatus*
  b. Cap shorter, bell shaped to conical, not white, having scales or particles mostly at or near center of cap 3

3a. Cap whitish gray to black, scaly or smooth 4
  b. Cap tan to reddish brown, glistening particles frequently covering center of cap . . . .*Coprinus micaceus*

4a. Cap smooth to scaly over the center only; usually on soil near stumps
    or from buried wood . . . .*Coprinus atramentarius*
  b. Cap at first covered with flakes or scales; usually on rotting wood . . .
    . . . .*Coprinus quadrifidus*

5a. Cap small, less than 1.5 cm broad; occurring in large numbers on well-
    decayed wood. . . . .*Coprinus disseminatus*
  b. Cap larger; single or in small groups 6

6a. Fruiting body without a partial veil . . . .*Panaeolus foenisecii*
  b. Fruiting body with a partial veil when young 7

7a. Cap buff fading to whitish; margin irregular. . . . .
    . . . .*Psathyrella candolleana*
  b. Cap tawny brown to yellowish brown; margin regular . . . .
    . . . .*Psathyrella velutina*

34 *Coprinus atramentarius*

35 *Coprinus comatus*

Coprinus atramentarius (Bull.: Fr.) Fr. **(34)**

"INKY CAP"

Edible, conditionally

Cap 4–6 cm broad to 6 cm high, conic becoming bell shaped at maturity; white to gray to black to inky with age; smooth to scaly over center; margin striate, wrinkled when young with a thin fibrillose partial veil. Flesh thin; pallid.

Gills free to adnexed, crowded; whitish dissolving into an inky black fluid with age.

Stalk 8–15 cm long, 1–1.5 cm thick; white above annulus zone, brown fibrils scattered below; hollow.

Annulus membranous, fibrous.

Spores 7–12 × 4.5–5.8 μm, elliptical, smooth; apical pore; spore print black.

Several in a group to dense clusters in grass, on wood debris, or from buried wood, Sp., S., F.

This species is reported to be **toxic** when ingested with **alcoholic** beverages. The toxic reaction involves nausea, a flushed appearance with a hot feeling, and gastrointestinal disturbance. Under other conditions it is a good edible species.

Coprinus comatus (Muller: Fr.) S. F. Gray **(35)**

"SHAGGY MANE" or "LAWYER'S WIG"

Edible, choice

Cap 5–12 cm tall, 2–6 cm across base just before spore maturation, egg shaped expanding to bell shaped; white with shaggy recurved scales tinged with brown; margin ribbed to wrinkled. Flesh soft; white darkening and dissolving with age.

Gills nearly free to notched and very crowded; white to pink dissolving to a black inky mass.

Stalk 5–18 cm long, 1–1.5 cm thick, equal, slightly bulbous at base; creamy white; dry.

Annulus fibrous, a movable ring that may slip to base of stalk or may disintegrate.

Spores 13–17 × 7–8 μm, elliptical, smooth; apical pore; spore print black.

Scattered or in small groups on hard ground or grass, along paths, roadsides, late S., F., sometimes Sp.; more common in F.

36 A, B: *Coprinus disseminatus*

Coprinus disseminatus
(Pers.: Fr.) S. F. Gray   **(36A, B)**

Nonpoisonous

Cap 0.6–1.5 cm broad, conic to bell shaped; grayish to gray-brown with center pale buff to honey brown; striate with folds to the margin. Flesh soft, thin; pallid.

Gills attached, subdistant; white to gray then black but not inky at maturity.

Stalk 2–4 cm long, 0.1–0.2 cm thick, equal, curved; white; smooth, hollow.

Spores 7–10 × 4–5 μm, elliptical, smooth; apical pore; spore print black.

In large numbers (hundreds) on well-decayed wood debris or buried wood, common on or around old stumps, Sp., S., F.

This mushroom has also been interpreted as a species of *Pseudocoprinus* because the gills do not deliquesce.

37   *Coprinus micaceus*

38   *Coprinus quadrifidus*

Coprinus micaceus (Bull.: Fr.) Fr. (37)

Edible

"MICA CAP"

Cap 1.5–5 cm broad, 2.5–4 cm high, conic to bell shaped at maturity; tan to reddish brown covered with glistening white to light tan particles often disappearing with age; strongly striate on edge and smooth at top. Flesh soft, watery; pallid.

Gills notched, crowded; white turning first gray then black at maturity dissolving into an inky fluid.

Stalk 1.5–7 cm long, 0.4–0.5 cm thick, equal; white; hollow.

Spores 7–10 × 3.5–5 μm elliptical, smooth; apical pore; spore print black.

In dense clusters on rotten wood, near stumps or buried wood, after rains, Sp., S., F.

Very common in large fruitings after rains in spring and fall, less abundant in summer. The mushrooms are delicate and soon deliquesce resulting in dark inky masses in a lawn.

Coprinus quadrifidus Pk. (38)

Edible

Cap 2–5 cm broad at base, 2–4 cm high, conic to bell shaped; cream to pale gray to grayish brown covered with whitish to dingy flakes or scales that may disappear with age; margin wavy and uneven. Flesh thin.

Gills reaching to the stalk but free, crowded, broad; whitish becoming purplish then black and inky.

Stalk 4–12 cm long, 0.5–0.8 cm thick, equal or tapering upward; whitish, sometimes with white cords of mycelium at base.

Annulus floccose, attached irregularly toward basal portion of stalk, often disappearing with age.

Spores 7.5–10 × 4–5 μm, elliptical, smooth; spore print black.

Common in large clusters on rotten logs or piles of woody debris, early S., F.

Coprinus atramentarius is similar in appearance but lacks the floccose patches on the cap. Also, C. quadrifidus occurs on rotten wood with well-developed rootlike cords of mycelium (rhizomorphs).

39 *Psathyrella candolleana*

Panaeolus foenisecii Kühner Poisonous, common

Cap 1–3 cm broad, convex to conic; brown to grayish brown to reddish brown to violaceous brown drying to lighter tan center and a darker margin; margin smooth to faintly striate. Flesh thin, watery; tan to brown.

Gills attached, adnate, close; often mottled then chocolate brown to purplish black, edges white.

Stalk 5–8 cm long, 0.2–0.4 cm thick, sometimes enlarging at base; pale tan at top, darker dingy brown below.

Spores 12–20 × 8–10 μm, broadly elliptical, roughened with low, irregular warts; apical pore; spore print dark purple-brown.

Scattered to numerous in lawns and pastures, Sp., S., F.

This mushroom may contain hallucinogenic compounds (psilocybin and psilocin) and has been reported to cause poisoning. It has also been called *Psathyrella foenisecii*. It is widely distributed and common in lawns, particularly in the spring and summer. The cap usually does not expand much with age. Many species of *Panaeolus* occur on dung, and there are also other species in grassy areas.

Psathyrella candolleana (Fr.) Maire **(39)** Edible, common

Cap 3–8 cm broad, rounded to finally upturned at maturity; buff to honey color fading to whitish, with scattered small white scales when young; veil remnants may be found hanging from margin; margin irregular. Flesh thin, fragile; white.

Gills attached, crowded; white to grayish maturing purplish brown.

Stalk 3–10 cm long, 0.2–0.5 cm thick, rigid; shiny white; smooth, hollow.

Partial veil membranous; white; annulus rarely forming; veil tissue remaining attached in separate pieces to young cap margin disappearing with age.

Spores 6.5–8 × 4–5 μm, elliptical, smooth; apical pore; spore print purple-brown.

Single to scattered in grassy lawns or forest openings, Sp., S., F.

This mushroom is also known as *Hypholoma incertum*. The irregular, radially asymmetrical margin of the cap is a distinctive field characteristic. The partial veil remnants tend to dry and are lost quickly, particularly in drying situations.

Psathyrella velutina (Fr.) Sing. (**40**)

Cap 3–8 cm broad, convex becoming plane with a slightly raised central area; tawny brown to yellowish brown with darker center; appressed fibrils becoming fibrillose scales; margin often fringed with remnants of partial veil, splitting. Flesh thick; brownish.

Gills attached to nearly free, close to crowded, broad; edges white and finely hairy with beads of moisture in wet weather becoming dark purple-brown.

Stalk 3–7 cm long, 0.4–1 cm thick, equal; whitish above ring, brownish below; hairy to scaly especially below ring, hollow with age.

Partial veil fibrous, breaking with remnants on edge of cap; annulus an obscure ring on stalk.

Spores 9–12 × 6–8 µm, elliptical to ovoid, warted; spore print dark purple-brown.

Single or in scattered groups of two or three, in grassy areas in lawns or in woods, S., F.

40 *Psathyrella velutina*

Cortinariaceae

This large family includes mushrooms that are either sessile or eccentrically stalked (only the genus *Crepidotus*) or centrally stalked. The caps are rounded to nearly flattened at maturity, yellow to shades of brown or purple, dry to viscid, with entire to slightly striate margins. Flesh colors are white to shades of brown and purple.

Gills are adnate to adnexed or sometimes decurrent, and range in color from tan or buff to purple or red to brown with color darkening or changing as spores mature.

Stalks are very thick, some bulbous at the base, smooth to hairy. Partial veils when present are cobwebby to floccose, often obscure and easily overlooked, or soon disappearing.

Spores are yellow-brown to rusty brown, round to elliptical, smooth to variously ornamented with knobs, spines, ridges, or wrinkles, and the spore prints are yellow-brown to rusty brown.

We have included only a few examples from this family despite the fact that many occur in this region. The largest genus, *Cortinarius*, is very difficult taxonomically. There are a few distinctive species, but many are not easily recognized or identified without careful microscopic study.

KEY TO SELECTED GENERA AND SPECIES IN THIS GUIDE

(Be aware of possible misidentification. Do NOT eat any specimen not positively identified.)

1a. On wood; stem eccentric, lateral or lacking. . . . . . .*Crepidotus mollis*
 b. On ground; stem present, usually central .2

2a. Veil membranous; cap orange with a wrinkled margin . . . .
 . . . .*Rozites caperata*
 b. Veil cobwebby, like a spider web, if present; cap not as above. 3

3a. Cap conic to bell shaped, typically radially fibrillose. 4
 b. Cap usually convex, not distinctly radially fibrillose . . . .
 . . . .*Cortinarius* spp.

4a. Cap sharply conic, yellow-brown . . . .*Inocybe fastigiata*
 b. Cap becoming almost flat with a knob, white . . . .*Inocybe geophylla*

Cortinarius spp. (41A-C)   Some poisonous species

Cortinarius is a large genus of mycorrhizal mushrooms with several hundred species and has been divided into several subgenera. Identification is very difficult, often requiring observation of a series of developmental stages and microscopic information. Thus, this genus is **dangerous** to the collector looking for edible fungi and should be **avoided**.

These mushrooms are usually quite fleshy, with convex caps, a cobwebby partial veil of silky hyphae when young, somewhat fleshy stalks, and rusty brown to cinnamon brown spores. Some have viscid caps or stalks, some do not, along with other varying characters. Many species are mycorrhizal, thus these mushrooms are generally found on the ground associated with trees. Some species are beautiful shades of purple when young, with the colors fading or remaining in different groups of species. They usually fruit in late summer and fall.

We have not attempted to identify the common species because of the difficulties of identification at the species level. The figures illustrate some aspects of the genus.

We again emphasize the danger involved with eating species of the genus *Cortinarius* even though some species are reported to be edible. Some species are **deadly** poisonous. *Cortinarius orellanus* is not only deadly but the symptoms may not appear for days or even weeks, and medical treatment may be very difficult. This species has not been reported from North America, but other species believed to contain the same or similar toxic compounds do occur. Do **NOT** eat mushrooms from the genus Cortinarius.

Crepidotus mollis Peck   Edibility unknown

Cap 1–5 (to 8) cm broad, 2–2.5 cm wide, convex; buff with small dark brown scales; dry; sessile or with rudimentary stalk. Flesh thin, soft; white.

Gills attached, close to subdistant, narrow; becoming cinnamon; cystidia on edges and sides of gills.

Stalk rudimentary if present.

Spores 7–10 × 4.5–6 μm, short elliptical, smooth; spore print yellow brown.

In overlapping groups on bark of dead hardwoods, sometimes on conifers, widely distributed, Sp., S., F.

> There are a number of species of *Crepidotus*, all with small, convex to fan-shaped, sessile caps, and found on wood. They are too thin to be considered edible. *C. applanatus* has 1–4 cm broad caps, is smooth to slightly downy, dull white to cinnamon, and is widely distributed on hardwood logs and stumps.

41 A, B, C: *Cortinarius* spp

Inocybe fastigiata (Schaeff.) Quél   **(42)**   Poisonous

Cap 2–6 cm broad, conic to campanulate with a central knob becoming expanded; tawny to yellowish brown, darker in center; dry, splitting readily on the margin, streaked with fibers. Flesh thin except at center; white.

Gills attached, notched at the stem, crowded; white becoming brownish.

Stalk 4–10 cm long, 2–8 mm thick, twisted striate, equal to slightly enlarged at base; whitish to light brown.

Spores 9–13 × 5–7 μm, elliptical, smooth; spore print dull brown.

Scattered or in groups under hardwoods or conifers and in grassy areas or lawns under trees, S. into F.

All species of *Inocybe* that have been studied have been found to contain toxins. Thus, all species are potentially poisonous and should **never** be eaten. *I. fastigiata* is one of the most common *Inocybe* species and is quite variable in size and color.

42   *Inocybe fastigiata*

Inocybe geophylla (Sow.: Fr.) Kummer   Poisonous

Cap 2–4 cm broad, rounded to bell shaped with a prominent pointed apex; white, silky fibrillose, dry; margin upturned with age. Flesh thin except at center; white.

Gills attached or notched at the stalk, crowded; white to grayish clay.

Stalk 2–7 cm long, 0.2–0.5 cm thick, solid; white to grayish white, silky fibrillose.

Partial veil cobwebby; white, annulus a faint hairy ring on stalk.

Spores 7–10 × 4.5–6 μm elliptical, smooth; spore print clay-brown.

Scattered under hardwoods and mixed woods, occasionally in lawns, S., F.

Rozites caperata (Fr.) Karst.   Edible

Cap 5–12 cm broad, rounded oval to bell shaped, wrinkled or grooved; tan to orange-brown, covered early with fine white hairs such that the surface appears frosted. Flesh thick, firm; white.

Gills attached, crowded; pale tan often streaked with brownish bands, rusty at maturity.

Stalk 6–10 cm long, 1–2 cm thick, smooth, enlarged toward base; dingy white.

Partial veil membranous; white; annulus persistent.

Spores 11–14 × 7–9 μm, elliptical, minutely warted; spore print rusty to yellowish brown.

Several to numerous in mixed woods or conifers, late S., F.

Gomphidiaceae

This family is characterized by a smoky gray to blackish spore print and thick decurrent gills. There is no clear distinction between cap and stalk because of the decurrent gill development. Probably all species are mycorrhizal with conifers; thus the mushrooms develop under or near conifers.

Caps may be viscid to tacky to dry, with or without a partial veil. This is a small family and while they are not reported to contain toxins, they are not particularly palatable because the flesh turns color upon cooking and the texture becomes somewhat slimy.

Gomphidius glutinosus (Schaeff.: Fr.) Fr. (43)

Cap 2–10 cm broad, broadly convex to flat; gray-brown to purple-gray or reddish brown often spotted or stained blackish; glutinous, smooth; margin upturned with age. Flesh thick, soft; white, sometimes pinkish.

Gills decurrent, close to subdistant; whitish then smoky gray to blackish.

Stalk 4–10 cm long, 1–2 cm thick, tapering toward base; white above slimy ring, yellow below ring.

Partial veil with thin, glutinous outer layer, white hairy inner layer; annulus a glutinous ring, darkens with accumulated spores with age.

Spores 15–21 × 4–7.5 μm, cyclindric, tapered at both ends, smooth; spore print smoky gray.

Single to scattered under conifers, late S., F.

43 *Gomphidius glutinosus*

Hygrophoraceae

This is an unusually colorful and conspicuous group. Cap colors range from reds, oranges, and yellows to white, gray, brown, and even green. Cap shape varies from conic to nearly flat, with the surface often viscid or slimy.

Gills are the distinctive feature of the group. They appear waxy and feel waxy when rubbed between the fingers, are thick (broadly triangular in cross section), well separated, and adnate to decurrent.

Stalks tend to be even, sometimes with a bulbous base, frequently cap color or slightly lighter often bruising to a darker color, and usually hollow with a tendency to split.

Spores are uniformly smooth, white, and elliptical; the spore prints are white.

In a more technical interpretation of the genus *Hygrophorus*, groups of species are characterized by the structure of the gill tissue. This microscopic character is valuable in identification, but the species discussed here can be interpreted without involving interpretation of gill tissue structure.

All species of *Hygrophorus* grow in litter or in the soil. The unusual waxy gills make this a relatively easy group to recognize. *Laccaria* species also have waxy gills, and the common mushroom, *Laccaria laccata,* may be mistaken for a *Hygrophorus* when first collected.

We include only a few distinctive species of *Hygrophorus.* Many more occur in this area.

KEY TO THE SPECIES OF *HYGROPHORUS* IN THIS GUIDE

(Be aware of possible misidentification. Do NOT eat any specimen not positively identified.)

1a. Cap whitish or some shade of red-orange or yellow-orange 2
  b. Cap grass green to bluish green when fresh . . . .*H. psitticinus*

2a. Cap red to scarlet-orange 3
  b. Cap white, pale yellow, pinkish, reddish yellow, to orange-brown . . .
 4

3a. Cap scarlet red to orange; gills and stalk bruising black . . .*H. conicus*
  b. Cap deep blood red aging to orange; gills and stalk not bruising to black . . . .*H. puniceus*

4a. Cap white, whitish yellow, reddish yellow aging to orange-buff, not changing color when bruised. 5
  b. Cap usually pinkish or pinkish brown; cap, gills and stalk bruising yellow when young . . . .*H. russula*

5a. Cap cream to tan with central area light orange-brown to apricot-buff,
    slimy to sticky; stalk slimy at base . . . .*H. subsalmonius*
  b. Cap pure white; surface slimy to sticky; stalk slimy .*H. eburneus*

Hygrophorus conicus (Fr.) Fr. **(44)**   Not recommended

Cap 2–7 cm broad, conic with an acute peak; red to scarlet-orange,
paler to olive near margin; viscid when wet; margin uneven, somewhat
upturned at maturity. Flesh thin; cap color bruising or aging to olive-gray to
black.

Gills nearly free, crowded, edges uneven; whitish yellow to olive bruising black; waxy.

Stalk 4–12 cm long, 0.3–1 cm thick, equal, striate-twisted; yellow to
scarlet but white at the lower part, black when bruised or when mature;
moist, hollow.

Spores 9–12 × 5–7 μm, elliptical, smooth; spore print white.

Scattered to somewhat gregarious, on ground, S., F.

This species is common in the loess hills of western Iowa and is not uncommon in woods. It is easily recognized because of the conic, brightly colored
cap and stalk, which quickly become black when bruised or broken.
    This species is sometimes designated *Hygrocybe conica*.

Hygrophorus eburneus (Fr.) Fr. **(45)**   Edible

Cap 2–7 cm broad, nearly rounded convex becoming knobbed or flat
to sunken; white to yellowish with a silky luster; slimy to sticky when wet,
often smooth; margin varying from inrolled to elevated at maturity. Flesh
thin; white to off-white.

Gills decurrent, nearly distant to distant, fairly broad, thicker near stalk;
white; waxy.

Stalk 5–15 cm long, 0.2–1 cm thick, uniform diameter to tapering
toward base; white; slimy.

Spores 6–9 x 3.5–5 μm, elliptical, smooth; spore print white.

Single to scattered in hardwoods or conifers and into adjacent grassy
areas, S., F.

This species is widely distributed but occurs in limited numbers. Distinguished by the very glutinous covering on both pileus and stalk, it is shiny
when dry.

44 *Hygrophorus conicus*

45 *Hygrophorus eburneus*

46   *Hygrophorus psittacinus*

47   *Hygrophorus puniceus*

Hygrophorus psittacinus (Fr.) Fr. **(46)** Edibility unknown

Cap 1–3.5 cm broad, conic or bell shaped to rounded; grass green to bluish green maturing to pinkish buff; viscid; margin striate. Flesh thin; cap color.

Gills adnate, moderately separated; light green becoming orange-yellow or buff, sometimes more evidently green near cap; waxy.

Stalk 3–7 cm long (about twice as long as cap width), 0.1–0.5 cm thick, equal; green becoming yellow-orange to cream; hollow, slimy.

Spores 6.5–9 × 4–5 μm, elliptical, smooth; spore print white.

Scattered in the soil under hardwoods and conifers, S. but mostly F.

This species is most common following heavy rains. Although it is widely distributed, it seldom occurs in large numbers. It is sometimes known as *Hygrocybe psittacina*. The unique green color is distinctive, though it disappears with age.

Hygrophorus puniceus (Fr.) Fr. **(47)** Edible

Cap 2–5 cm broad, rounded to bluntly conic; deep blood red aging to orange; viscid; margin incurved when young, somewhat striate. Flesh thin, watery; red-orange to yellow.

Gills attached, broad, fairly well separated; red-orange to yellow; waxy.

Stalk 3–7 cm long, 0.5–1.5 cm thick, equal or narrowing slightly at the base; reddish fading to orange-yellow, base white-yellow; fibrillose-striate; center hollow to filled.

Spores 8–11 × 3.5–6 μm, elliptical to oblong, smooth; spore print white.

Scattered to groups on soil under hardwoods and conifers, S., F.

Hygrophorus coccineus is similar and occurs in same kind of habitat. It differs in having a moist to tacky, but not truly viscid, cap and is reported to be less common. There are also other small, red to orange species of *Hygrophorus*.

48 *Hygrophorus russula*

49 *Hygrophorus subsalmonius*

Hygrophorus russula (Fr.) Quél. **(48)** Edible

Cap 5–12 cm broad, rounded, convex maturing to nearly flat; pink to pinkish brown or whitish, streaked with purplish fibrils or flecked with wine-colored spots, sometimes bruising yellow when young; smooth at first, breaking into patches of fibrils. Flesh firm; white to pinkish.

Gills attached or extending slightly down the stalk, crowded, firm; white but soon pale pink staining spotted purplish red at maturity; waxy.

Stalk 3–7 cm long, 1.5–3.5 cm thick, equal; white when young staining pink to reddish brown at maturity; dry, solid.

Spores 6–8 × 3–5 μm, elliptical, smooth; spore print white.

Scattered or clustered under hardwoods, common under oak, late S., F.

This most common fall mushroom in some years, particularly in rainy falls, has the appearance of a species of *Russula*. Sometimes fruiting occurs in arcs or in fairy rings. The firm, waxy gills are a good field characteristic.

Hygrophorus subsalmonius Sm. & Hesler **(49)** Edibility unknown

Cap 4–12 cm broad, rounded to convex becoming plane to slightly depressed with age; cream to tan with central area light orange-brown to apricot-buff, uniformly colored at maturity; slimy to sticky (glutinous to viscid) all over, smooth except for the cottony margin. Flesh thick, firm; white.

Gills attached, usually somewhat decurrent, extending slightly along stalk, close to subdistant, moderately broad, alternately long and shorter; light buff to pinkish buff; waxy.

Stalk 3–10 cm long, 0.5–2.0 cm thick, same width throughout or tapering slightly at base; whitish to tan darkening with age; lower portion slimy to sticky, solid.

Spores 6.5–8 × 3–4.5 μm, elliptical smooth; spore print white.

Scattered or clustered under oaks or hickory in woods, F. into late F.

A relatively common late fall mushroom, but one that is often not seen because it is under the cover of recently fallen leaves.

Lepiotaceae

This family is characterized by caps that are rounded to nearly flat at maturity, scaly to smooth with the scales part of the cap tissue not remnants of a universal veil. Cap margins are entire, fringed, or striate. The flesh is white, thick, and soft to firm, unchanging or bruising yellow or reddish brown. Partial veil is present when young, persisting as an annulus when mature.

Gills are free, close to crowded, and white to light tan or greenish.

Stalks are somewhat enlarged toward the base, smooth to hairy scaly, and have a prominent annulus, which becomes loose on the stalk in the larger species. The lack of a volva in the Lepiotaceae clearly distinguishes this group from the Amanitaceae.

Spores vary from elliptical to ovoid and the spore prints are white in all except *Chlorophyllum molybdites,* which has green spores and spore print.

This family presents problems to mushroom collectors: one is its resemblance to species of *Amanita;* another is that some small species of the Lepiotaceae are reported to contain alpha-amanitin and thus have the potential for causing severe **illness** or **death**.

KEY TO THE GENERA AND SPECIES IN THIS GUIDE

(Be aware of possible misidentification. Do NOT eat any specimen not positively identified.)

1a. Spore print green; gills green at maturity . . . .
. . . .*Chlorophyllum molybdites*
  b. Spore print white 2

2a. Cap surface smooth. . . . .*Leucoagaricus naucinus*
  b. Cap surface with scales, which do not rub off easily. 3

3a. Cap large (6–24 cm); stalk long (12–30 cm) 4
  b. Cap smaller (2–5 cm); stalk shorter (2.5–7 cm) 5

4a. Cap flesh and stalk unchanging when bruised or cut. . . . .
. . . .*Lepiota procera*
  b. Cap flesh and stalk bruising orange-red to red-brown . . . .
. . . .*Lepiota rachodes*

5a. Cap dull reddish brown; reddish brown scales exposing whitish
  flesh . . . .*Lepiota cristata*
  b. Cap lemon yellow to cream-yellow; surface powdery or scaly yel-
  low. . . . .*Lepiota lutea*

Chlorophyllum molybdites (Meyer) Massee **(50)**

Poisonous

Cap 10–30 cm broad, hemispheric maturing to nearly flat, often with a broad central raised area; white to tan to cinnamon brown, cracking to form patchy scales above white flesh; dry. Flesh firm; white.

Gills free, crowded; white maturing to green bruising yellow-brown.

Stalk 5–25 cm long, 1–2.5 cm thick, smooth; white staining brownish especially toward enlarged base.

Annulus thick, firm; white discoloring brown beneath; may become movable.

Spores 9–12 × 6.5–8 μm, elliptical, thick walled; apical pore; spore print pale dull green, which distinguishes it from the edible *Lepiota procera.*

Scattered or forming fairy rings in lawns or pastures especially following heavy rains, S., early F.

A fairy ring is a striking phenomenon in itself but closer inspection reveals another interesting characteristic. A bank of lighter green grass is found outside the ring of mushrooms and a ring of lush, dark green grass occurs inside the ring. The mycelium of the fungus is growing vigorously in advance of the ring of fruiting bodies, using soil nitrogen and other nutrients in its assimilation of cellulose. Behind the ring the mycelium is dying and the nitrogen and other nutrients are released. They are quickly used by the grass.

Because this mushroom may be mistaken for edible species such as *Lepiota procera,* spore prints should be made before eating any mushroom of this type. The green spore print will distinguish *C. molybdites.* This mushroom is not poisonous to all people, but causes **severe** illness in some.

This species has been known as *Lepiota morgani* or *Lepiota molybdites* and is very common in moist years, producing spectacular fruitings.

50 *Chlorophyllum molybdites*

51 *Lepiota cristata*

52 *Lepiota lutea*

Lepiota cristata (Fr.) Kummer **(51)**   Poisonous

Cap 2–5 cm broad, rounded oval maturing nearly flat with central knob; dull reddish to reddish brown when young breaking into small reddish brown scales exposing the whitish surface underneath, central knob remaining reddish brown and smooth; scales often in concentric circles and finer at the edge; margin striate. Flesh thin; white.

Gills free, crowded; white; edges slightly wavy.

Stalk 3–5 cm long, 0.3–0.7 cm thick, equal; whitish or dingy pinkish to brownish at base; smooth or slightly downy.

Annulus fragile, often quickly breaking and disappearing; white.

Spores 6–8 × 3–4 μm, tapered on one end (wedge shaped); spore print white.

Several to scattered in woods or grassy areas, S., F.

Most common during summer in years of plentiful rainfall. They are small enough to escape notice. Some small species of *Lepiota* are **potentially deadly** poisonous and should be **avoided**.

Lepiota lutea (Fr.) Loa. **(52)**   Poisonous

Cap 3–6 cm broad, bell shaped becoming conical to nearly flat with a central knob; lemon yellow fading to cream-yellow; surface scaly or powdery; margin deeply striate. Flesh very thin; light yellow.

Gills free, crowded, thin; yellowish cream with hairy edges.

Stalk 3–8 cm long, 0.15–0.5 cm thick, often enlarged toward the base; usually with evident powdery yellow scales.

Annulus lemon yellow, readily breaking down.

Spores 8–13 × 5–8 μm, elliptical, smooth; spore print white.

Several to numerous on the ground in leaf litter or compost, potting soil, and greenhouse benches, S., F. outdoors, or any season in greenhouse and house plant soils.

We know this species as a mushroom occurring with potted plants in greenhouse beds, but it is a widely distributed field species in the southern United States in summer.

This has been known as *Leucocoprinus birnbaumii* and *Leucocoprinus luteus*. *L. lutea* and many small species of *Lepiota* are reported to be poisonous.

53 *Lepiota rachodes*

Lepiota procera (Scopoli: Fr.) S. F. Gray   Edible
"PARASOL MUSHROOM"

Cap 6–24 cm broad, ovoid expanding to broadly bell shaped with small knob at center; gray-brown to reddish brown surface breaking into scales exposing whitish flesh. Flesh thick, soft; white not changing color when bruised.

Gills free, close, crowded; white; hairy edges.

Stalk 15–30 cm long, 1–1.5 cm thick, enlarging to a bulbous base; gray-brown, irregular brown scaly markings and cracks showing white flesh; felty; fine, tan hairs above the annulus.

Annulus thick, soft; white; easily detaching and movable.

Spores 12–15 × 7.5–10 μm, ovoid; small apical pore; spore print white.

Several to scattered in brush, lawns, or woods, late S., F.

This species can be easily confused either with species of *Amanita* or *Chlorophyllum*. The scales on the cap of *L. procera* are actually part of the cap and not superficial as are cap scales of *Amanita* species. The spores of *Chlorophyllum molybdites* are green while those of *Lepiota procera* are white.

Lepiota rachodes (Vitt.) Quél. (53)   Not recommended

Cap 6–20 cm broad, dry, conic to rounded to nearly flat at maturity; gray-brown to reddish brown fracturing to form concentric rings of up-turned scales exposing lighter flesh beneath; low central knob smooth, brownish. Flesh whitish (or pinkish at cuticle surface) bruising orange-red but soon becoming red-brown.

Gills free, close; whitish buff bruising to yellowish brown or darker brown.

Stalk 10–15 cm long, 1.5–3 cm thick, club shaped enlarging to a bulbous base; white above annulus, grayish white below with gray-brown hairs above bulb; surface bruising red-brown; flesh bruising as cap flesh.

Annulus soft, thick, two layered; white; becoming movable.

Spores 8–10 × 6–7 μm, broadly elliptical; spore print white.

Single to clustered in lawns, leaf and compost piles, in open areas or under trees, often forming fairy rings, late S., F.

Because of possible confusion with *Chlorophyllum molybdites*, this species **cannot** be recommended for consumption.

Leucoagaricus naucinus (Fr.) Sing. **(54)** Not recommended

Cap 4–8 cm broad, ovoid maturing to rounded and finally flattened; dull white or rarely gray, finally tannish; dry, smooth to sometimes minutely scaly. Flesh thick, soft; white.

Gills free, crowded; white maturing pinkish gray; rounded near stalk.

Stalk 5–12 cm long, 1–2 cm thick, nearly equal, enlarged to a bulbous base; cap color; smooth above the annulus, silky below.

Annulus membranous, thick; white; stiff and collarlike.

Spores 7–9 × 5–6 μm, ovoid or nearly so, thick walled; small apical pore; spore print usually white, sometimes pinkish.

Scattered in lawns, pastures, and occasionally under hardwoods, especially after heavy rains, S., F.

Because this species can be easily mistaken for the white Amanitas, we do **not** recommend this species. In addition, some forms of this species have been reported to cause gastric disturbances.

This species is also known as *Lepiota naucina*.

54 *Leucoagaricus naucinus*

Rhodophyllaceae (Entolomataceae)

This family has caps that are conic to pointed or rounded to broadly rounded or even flat at maturity. Cap colors range from white to dull grays and browns or pink- to orange-salmon. The cap flesh is soft and white to pallid off-white. Cap surfaces range from dry to tacky when wet, often drying to silky smooth, with margins frequently inrolled and splitting at maturity to entire to wavy.

Gills are attached and either notched or slightly decurrent.

Stalks are thin to moderately thick, nearly equal to somewhat enlarged at the base, often striate, brittle, and appearing somewhat twisted.

Spores are pink to brownish pink and either have long ridges or are angular.

This is typically a cool, rainy weather group in the Midwest including numerous species, but because characteristics of spore morphology are generally used to delimit genera and species, the group is not well known to amateurs.

We do **not** recommend this group for eating because of difficulties of identification, and because the toxins in the poisonous species are apparently varied with variable effects.

KEY TO THE GENERA AND SPECIES IN THIS GUIDE
(Be aware of misidentification. Do NOT eat any specimen not positively identified.)

1a. Gills decurrent; spores longitudinally striate . . . . .*Clitopilus prunulus*
 b. Gills attached, adnexed to short decurrent; spores angular 2

2a. Gills short decurrent; often normal mushrooms accompanied by oval, white modified fruiting bodies . . . .*Entoloma abortivum*
 b. Gills attached or adnexed; normal mushrooms 3

3a. Cap 8–15 cm broad; spore print salmon-pink . . . .*Entoloma sinuatum*
 b. Cap 3–7 cm broad; spore print deep rose. . . .*Entoloma rhodopolium*

55 *Entoloma abortivum*

Clitopilus prunulus (Fr.) Kummer   Not recommended

Cap 3–10 cm broad, convex to flat to shallowly depressed in center; white to grayish to yellowish; tacky when moist; margin inrolled to even to wavy. Flesh firm; white.

Gills decurrent, close to subdistant, narrow; white becoming pinkish.

Stalk 3–8 cm long, 4–15 mm thick, nearly equal; dull white; solid, usually central but occasionally slightly eccentric.

Spores 9–12 × 5–6.5 μm, tapering at ends, longitudinal ridges, pinkish; spore print salmon-pink.

Scattered to loosely grouped on the ground in open woods, both hardwood and coniferous, S., F.

Sometimes two species, *C. prunulus* and *C. orcellus* are recognized, separated primarily on the closeness of the gills of *C. orcellus*.

Entoloma abortivum (Berk. & Curt.) Donk   **(55)**   Edible
"ABORTED ENTOLOMA"

Cap 5–10 cm broad, rounded to nearly flat at maturity, somewhat knobbed at center; gray-brown; obscurely zoned with fibrous hairs to scales becoming smooth at maturity; margin inrolled. Flesh soft; white. Abortive, parasitized fruiting bodies are whitish, bumpy, soft, irregular masses occurring alone or near normal mushrooms.

Gills vary from merely attached to somewhat decurrent, somewhat crowded; gray aging to dusty rose or salmon-pink.

Stalk 4–10 cm tall, 0.8–2 cm thick, white to gray with scruffy hairy surface and white mycelium at the base.

Spores 8–10 × 4–6 μm, angular-elliptic; spore print tan-cinnamon to salmon-pink.

Scattered to frequent in hardwoods or mixed woods, frequently clustered at the base of red elms in moist habitats, around stumps and rotted wood, late S., F.

Parasitized mushrooms are typically highly modified, even looking like irregular lumps of soggy off-white tissue. There may be no evidence of either gill or stalk development.

This mushroom has been known under several names; currently, it may also be called *Rhodophyllus abortivus.*

While edible and valued by those who recognize this species, we urge **caution** because it is relatively easy to confuse this mushroom with poisonous *Entoloma* species.

56 *Entoloma rhodopolium*

Entoloma rhodopolium (Fr.) Kummer **(56)**

Not recommended

Cap 3–7 cm broad, convex becoming expanded to nearly flat; light brown to smoky brown when moist appearing water soaked, pale brownish gray when dry; surface smooth, slightly slippery. Flesh whitish; splitting easily.

Gills adnate becoming notched near the stalk, moderately widely spaced; whitish at first becoming deep rose.

Stalk 4–8 cm long, 0.5–1.0 cm thick, equal or slightly tapering, sometimes curved; whitish; smooth, easily splitting longitudinally.

Spores 8–10.5 × 7–9 μm, angular; spore print pink.

Solitary or in small clusters on the ground under hardwoods or in mixed woods, S., F.

There are a number of species of *Entoloma* with this somewhat general appearance developing at various times during the summer and fall season. All are poorly known and should be **avoided**.

Entoloma sinuatum (Bull.: Fr.) Kummer

Poisonous

Cap 8–15 cm broad, bell shaped to nearly flat with a small knob at maturity; dingy brown to tan; margin inrolled, maturing to straight, often split. Flesh thick; white.

Gills notched at stalk, fairly well separated; cream to gray-pink.

Stalk 4–12 cm tall, 1–4 cm thick, equal; white tinted grayish with scattered scales; dry, shiny, sturdy.

Spores 7–10 μm, nearly round but angular; spore print salmon-pink.

Scattered under hardwoods and mixed woods, S., F.

Russulaceae

The two genera in this family, *Russula* and *Lactarius,* are characterized by rounded to depressed caps and gills composed of tissue with groups of large round cells intermixed with regular cellular hyphae. Cap margins vary from inrolled, entire, or striate to split. Flesh is usually brittle and dry. *Lactarius* contains latex-producing cells with latex apparent when the gills or other tissue of young mushrooms are cut. Color changes in the latex are important in identifying species of *Lactarius. Russula* lacks latex cells, but gill and cap flesh may change color following cutting or bruising.

Gills of *Russula* and *Lactarius* are adnate (attached) or decurrent (extending down the stalk), range from close to distant, and are white to cream to yellow in color. Gills of *Russula* species break and crumble easily.

Stalks tend to be rather thick, equal, short, and often smooth and chalky or brittle in texture.

Spores are white to yellow, elliptical to nearly round, and roughened with warts and ridges developed in the spore wall. These wall ornamentations exhibit a blue amyloid reaction in Melzer's solution. Spore prints are white to cream to yellow.

There are edible as well as poisonous mushrooms in this group. It is **not** a group to experiment with. Species such as *Lactarius deliciosus* often fruit in large numbers and are among the favorites of mushroom hunters who know them. Many of the species are found associated with specific hardwood or conifer trees, and most of this group are probably mycorrhizal associates.

Characteristic field aspects of these mushrooms are their short stalks; broad, often brightly colored caps; and gills, particularly those of *Russula* species, that break and crumble easily.

KEY TO THE GENERA IN THIS GUIDE

1a. Cap exuding latex when cut or broken (*Lactarius*). 2
 b. Cap not exuding latex when cut or broken (*Russula*) 9

KEY TO THE SPECIES OF *LACTARIUS* IN THIS GUIDE

(Be aware of possible misidentification. Do NOT eat any specimen not positively identified.)

2a. Latex initially white 3
 b. Latex initially colored. 8

3a. Latex unchanging to slowly drying yellow or staining gills yellowish
. . . .*L. piperatus*
 b. Latex changing color staining gills other than yellowish 4

4a. Latex changing, slowly staining gills brown or pink to rose. 5
 b. Latex changing, slowly staining gills dingy gray-green. . . . .
. . . .*L. argillaceifolius*

5a. Latex slowly changing brown. 7
 b. Latex slowly changing, staining gills pink to rose or lilac 6

6a. Latex slowly changing, staining gills pink to reddish cinnamon; cap
chocolate-brown to black-brown . . . .*L. lignyotus*
 b. Latex changing white to cream-gray, staining gills lilac-tan; cap lilac
to violet . . . .*L. uvidus*

7a. Cap minutely velvety woolly, dull white to creamy tan . . . .
. . . .*L. subvellereus*
 b. Cap dry, smooth, buff to orange-brown . . . .*L. volemus*

8a. Latex bright blue; cap zoned blue . . . .*L. indigo*
 b. Latex carrot orange staining flesh green; cap carrot orange, zoned . .
. . . .*L. deliciosus*

KEY TO THE SPECIES OF *RUSSULA* IN THIS GUIDE
(Be aware of possible misidentification. Do NOT eat any specimen not positively identified.)

9a. Cap bright red or red-orange. . . . .*R. emetica* group
 b. Cap not as above, colored otherwise or white. 10

10a. Cap yellowish to brownish yellow, very striate . . . .*R. foetens* group
 b. Cap other than yellowish. 11

11a. Cap reddish purple often mixed with green or olive . . . . .*R. variata*
 b. Cap never with reddish purple tints. 12

12a. Cap green to grayish green. . . . .*R. aeruginea*
 b. Cap not green but whitish 13

13a. Cap color white staining dull yellow to brownish; flesh brittle, white
after bruising. . . . .*R. brevipes*
 b. Cap color white aging brownish black; flesh white slowly bruising
reddish. . . . .*R. nigricans*

57 *Lactarius argillaceifolius*

58 *Lactarius deliciosus*

Lactarius argillaceifolius
Not recommended

Hesler & Smith var. *argillaceifolius* **(57)**

Cap 5–18 cm broad, rounded maturing to flattened and finally depressed; pink to lavender-gray or yellowish brown to cinnamon; surface slimy, viscid, and glabrous; margin incurved, downy to smooth or obscurely zonate. Flesh thick, firm; white; latex white slowly staining dingy gray-green.

Gills adnate, close to subdistant; pallid or pale tan.

Stalk 1–2.5 cm thick × 5–10 cm long, equal; tan to pale cinnamon; often hollow, slimy turning dry.

Spores 6–7.5 × 7.5–9 μm, broadly elliptical; ornamentation an interrupted network to warty; spore print cream or dull yellow.

Solitary to gregarious in hardwood forests, late S.

This mushroom is usually present and is common in some years. It has been known as *L. trivialis* in North America and is a variable species.

Lactarius deliciosus (L.: Fr.) S. F. Gray **(58)**
Edible

Cap 5–15 cm, rounded maturing to flat or funnel shaped; carrot orange with concentric rings of darker orange and varied amounts of dull green, zoned with three to six distinctly orange-rust zones 1–3 mm wide; viscid to sticky in wet weather; margin incurved. Flesh pale yellow-orange staining green; latex carrot orange leaving green stains on flesh.

Gills crowded, extending down the stalk; bright orange, slowly becoming green when bruised.

Stalk 3–8 cm long, 1.5–3 cm thick, narrowed at the base; paler than cap bruising green when handled or at maturity; dry.

Spores 8–11 × 7–9 μm, elliptical to nearly round, warts and thickened bands on spore walls; spore print buff.

Single to scattered in coniferous woods, especially common under pines, late S., F.

There are several recognized varieties.

59 *Lactarius indigo*

Lactarius indigo (Schw.) Fr. **(59)**   Edible

Cap 5–15 cm broad, rounded to depressed in center; dark blue, usually zoned fading to pale blue-gray; surface sticky to smooth; margin inrolled at first. Flesh white bruising blue or becoming blue when cut, then turning blue-green to dark green with exposure; latex scant, dark blue turning to blue-green or dark green.

Gills attached, close; dark blue aging to pale blue-gray.

Stalk 2–8 cm long, 1–2.5 cm thick, often tapered; dark blue to silver-blue; hollow when mature, dry.

Spores 7–9 × 5.5–7.5 μm, broad elliptical to round, amyloid; spore print cream.

Scattered on soil in hardwood or coniferous woods, S., F.

This bright indigo blue *Lactarius* is usually a fall species in our region.

Lactarius lignyotus Fr.   Nonpoisonous

Cap 2–10 cm broad, rounded to broadly knobbed; dark chocolate brown to sooty black-brown; dry, velvety to somewhat wrinkled. Flesh brittle; white staining pink to rose when bruised; latex watery, white staining flesh-pink to reddish cinnamon.

Gills attached or slightly decurrent, subdistant; white bruising reddish, edges darker tan to brown near the stalk.

Stalk 4–12 cm long, 0.5–1.5 cm thick, enlarged near base; cap color; dry, furrowed at apex to smooth and velvety below.

Spores 9–11 × 8–10 μm, nearly round; reticulate ridges, amyloid; spore print buff-tan.

Several to grouped under white pine or mixed conifer woods, late S., F.

60 *Lactarius piperatus*

61 *Lactarius subvellereus*

Lactarius piperatus (Fr.) S. F. Gray **(60)**   Poisonous?

Cap 5–15 cm broad, convex becoming flat to finally depressed, funnel shaped; white staining creamy white to dingy tan with age; surface dry, smooth to slightly uneven; margin even. Flesh white unchanging to sometimes yellowish; latex white unchanging to slowly drying yellow or staining gills yellowish; taste very peppery.

Gills attached, crowded, narrow; white becoming pale cream; many of them forked one or more times.

Stalk 2–8 cm long, 1–2.5 cm thick, not tapering; white; dry, solid.

Spores 5–7 × 5–6 μm, roundish; warts and ridges amyloid; spore print white.

Scattered to clustered on soil under hardwoods, S., F.

The very crowded, frequently forked gills are a most important identification characteristic for this species.

Lactarius subvellereus Pk. var. *subdistans* Hesl. & Sm. **(61)**   Not recommended

Cap 5–15 cm broad, rounded to flattened to depressed; dull white to creamy tan; dry, minutely velvety woolly; margin inrolled at first but soon flattened. Flesh firm, thick; white; latex cream-white turning slowly brownish.

Gills short, decurrent, crowded; white to creamy tan slowly staining brown when cut.

Stalk 2–6.5 cm long, 1.5–3 cm thick, equal; dull white to cream; dry with minute velvety surface.

Spores 7–9 × 5–6.5 μm, elliptical; surface warty, amyloid; spore print white.

Single to scattered in hardwood or mixed forests, particularly under oaks, S., F.

This species has been known as *L. vellereus* in the past, but it is now evident that the smaller-spored *L. subvellereus* is separate. There are some people who eat this mushroom, but to others it is definitely toxic, thus **cannot** be recommended. There are other similar species with inrolled caps with a cottony roll at the margin.

62 *Lactarius uvidus*

63 *Lactarius volemus*

Lactarius uvidus (Fr.) Fr. **(62)** Poisonous

Cap 3–10 cm broad, rounded to flat, finally depressed; pale lilac darkening, unzoned; margin inrolled initially, smooth, sticky to dry. Flesh whitish gray; latex white turning cream-gray staining fleshy dull lilac.

Gills attached, close; cream becoming dull lilac to tan when bruised.

Stalk 3–7 cm long, 1–1.5 cm thick; whitish, often stained ochre-yellow at the base; sticky to shiny dry.

Spores 7.5–9.5 × 6.5–7.5 μm, rounded to broadly elliptical, amyloid; spore print pale yellow.

Single to scattered under pine, mixed hardwood-conifer woods, S., F.

Lactarius volemus (Fr.: Fr.) Fr. **(63)** Edible

Cap 5–10 cm broad, rounded becoming depressed; rusty tawny to rusty orange-buff to orange-brown; dry, smooth. Flesh white to dull white slowly bruising dark brown; latex white slowly changing brownish.

Gills extending down the stalk, crowded; cap color or lighter bruising brownish.

Stalk 5–8 cm long, 1–2 cm thick; cap color or lighter staining darker than cap.

Spores 7–10 × 7.5–8.5 μm, nearly round, ridged; spore print white.

Scattered on soil under hardwoods, S. Not uncommon during hot wet weather, S., early F.

A similar species, *L. corrugis,* differs in having a more distinctly velvety dark brown cap with a corrugated surface and somewhat larger spores. Both species occur primarily in deciduous woods.

64 *Russula aeruginea*

65 *Russula brevipes*

Russula aeruginea Lindbl. **(64)**   Inedible

Cap 3–8 cm broad, broadly rounded becoming depressed in the center; dull green to darker green or smoky green with a darker center and sometimes paler brownish margin; viscid when wet, slightly velvety when dry; margin slightly striate with age. Flesh thick in center, thinner at margin; white or greenish under surface.

Gills narrowly attached to nearly free, close to subdistant, equal or with a few short ones; white becoming cream.

Stalk 4–6 cm long, 1–2 cm thick, nearly equal; white; smooth.

Spores 7–9 × 5.5–7 μm, subglobose; white ornamented with small separate warts and a few small lines; spore print white to cream.

Solitary or grouped on the ground in coniferous or mixed woods, S., F.

Another green-capped species, *R. virescens,* also occurs in the Midwest. *R. virescens* has a green to grayish green cap with the dry surface soon breaking into patches exposing the white flesh. It is usually present and very common in some seasons.

Russula brevipes Pk. **(65)**   Inedible

Cap 10–20 cm broad, broadly rounded with depressed center becoming deeply depressed; white staining dull yellow to brownish; surface dry, minutely woolly; entire margin inrolled. Flesh brittle; white.

Gills attached to slightly decurrent, close to crowded, alternately long and short; white staining brownish.

Stalk 2.5–5 cm long, 2.5–4 cm thick; dull white staining brownish; dry, smooth.

Spores 8–11 × 6.5–10 μm, broadly elliptical, amyloid, ornamented; spore print white to cream.

Scattered on the ground under hardwoods or conifers, S., F.

This species may have been reported as *R. delica,* however, *R. delica* has more widely spaced gills and apparently does not occur here. The mushrooms are often partially covered by the soil and leaf litter through which they emerge. Some may remain concealed and rather large groups can be easily missed by collectors.

A

B

66   A, B: *Russula emetica*

Russula emetica group   **(66A, B)**   Poisonous?

Cap 5–10 cm broad, rounded becoming depressed; red to red-orange fading with age or damp weather; cuticle when peeled leaves a pink stain; margin striate, upturned in maturity. Flesh soft, fragile; white, pink just under cap surface; very hot peppery taste.

Gills attached, crowded; white; sometimes forked.

Stalk 5–10 cm long, 2.5–3 cm thick, usually longer than cap width; dull white; equal, often ridged, fragile, hollow at maturity.

Spores 8–10 × 6–8 μm, round or nearly so; warty ridges; spore print yellow.

Singly or in groups on soil in deep moss or near rotting conifers, S., F.

What we have described here probably includes a number of species constituting the *Russula emetica* group. They should be **avoided** for eating, but are often colorful and impressive in the woods.

67 *Russula foetens*

Russula foetens Fr. group (67)   Not recommended

Cap 5–14 cm broad, broadly convex to slightly depressed; yellowish to brownish yellow; sticky to viscid in wet weather, smooth; margin tuberculate-striate; odor becoming stronger and fetid with age. Flesh thin, rather fragile; dingy white.

Gills adnexed, rather close, broad; white becoming yellowish with age and bruising slightly darker, exuding drops of water when young; sometimes forked.

Stalk 3–6 cm long, 1–2.5 cm thick, equal; dull white becoming brownish with age; smooth.

Spores 8.5–10 × 8–9 μm, almost round with coarse projections; spore print white.

Scattered or in clusters on the ground under hardwoods or in mixed woods, S.

There is a group of species that is similar to the species described here as *R. foetens*. This may not be the appropriate name for this mushroom, but it has been widely reported under this name. The complex of similar species vary in various aspects, but all have yellowish to yellow-brown strongly striate caps with disagreeable taste and odor. They should all be **avoided** but are not likely to be eaten because of the unpleasant odor.

Russula nigricans (Bull.) Fr.   Not recommended

Cap 10–20 cm broad, rounded to flattened or depressed; whitish becoming smoky to brownish black; viscid when moist; margin incurved. Flesh white, slowly turning reddish upon bruising or cutting then becoming black.

Gills adnexed, unequal alternating long and short, broad, fairly well separated; white becoming reddish then black when bruised.

Stalk 3–6 cm long, 1.5–2.5 cm thick, equal; whitish becoming smoky brown with age changing reddish to black when bruised.

Spores 7.5–10 × 6–8 μm, nearly round, wall with low warts joined by a network of fine lines; spore print white.

Single to scattered in woods or openings, late S.

Despite its unattractive appearance as it matures, this is considered by some to be an edible mushroom of value.

R. nigricans is one of a small group of *Russula* species with alternate long and short gills and flesh becoming blackish with age or with bruising. Old fruiting bodies of *R. nigricans* may be completely blackened. *R. densifolia* differs from *R. nigricans* principally by the close to crowded gills. These species may have been confused, and there is some confusion about their frequency of occurrence and their habitats.

R. nigricans is often parasitized by *Asterophora lycoperdoides*. The parasite produces a number of small mushrooms from the cap or occasionally from the stalk of the parasitized specimen.

Russula variata Banning   **(68)**   Edible

Cap 5–15 cm, rounded to depressed in center; reddish purple to brownish purple often mixed with olive or green, sometimes quite greenish; viscid and shiny becoming dry; margin cracking with age. Flesh firm; white to grayish just under surface.

Gills adnate to slightly decurrent, close to crowded, narrow; white; forking two or three times, usually some distance from stalk.

Stalk 3–9 × 1–3 cm, dull; whitish, or stained brownish at base.

Spores 7.5–10 × 6–8.5 μm, subglobose; ornamented with low warts; spore print white.

Often common, scattered to loosely clustered on ground under hardwoods, early S., early F.

The confusing mix of cap colors and the narrow forking gills are distinguishing field characteristics.

68 *Russula variata*

Strophariaceae

This family includes mushrooms with spore-print colors ranging from purple-brown to yellow-brown. The fruiting bodies somewhat resemble those of the Coprinaceae but are more pliant and tend to be of brighter colors. *Pholiota* has often been placed with the Cortinariaceae and is included here because of the submicroscopic apical pore, which is a spore characteristic of all other genera in the Strophariaceae.

Caps range in color from bright yellow to tints of green or brown and are usually viscid or scaly, rounded to flattened at maturity, with margins often scaly or hairy. Gills are adnate close to well separated, and yellowish to darker with spores at maturity.

Stalks vary from thin to thick, smooth to hairy or scaly, with or without an annulus, and some with a slightly enlarged base.

Spores are smooth, usually more or less elliptical with an apical pore; spore prints are yellow-brown or rust-brown to purple-brown.

Several species of *Stropharia, Hypholoma,* and *Pholiota* are poisonous. Some *Psilocybe* species are reported to be hallucinogenic when consumed in moderate amounts but definitely toxic or fatal in larger amounts. Exercise **caution** when eating any members of this family.

KEY TO THE GENERA AND SPECIES IN THIS GUIDE

(Be aware of possible misidentification. Do NOT eat any specimen not positively identified.)

1a. Annulus well developed and evident. 2
  b. Annulus absent or thin and fibrillar 3

2a. Cap viscid, yellow; growing on dung. . . . .*Stropharia semiglobata*
  b. Cap viscid, buff to yellow-brown; growing in grass. . . . .
  . . . .*Stropharia coronilla*

3a. Cap margin buff to cinnamon, partial veil remnants inconspicuous . .
  . . . .*Naematoloma sublateritium*
  b. Cap margin yellow-brown, scaly; margin wavy from the partial veil remnants. 4

4a. Partial veil remnants cottony, white on cap and stalk*Pholiota destruens*
  b. Partial veil remnants scaly, yellow to greenish brown .*Pholiota squarrosa*

69 *Naematoloma sublateritium*

70 *Pholiota destruens*

Naematoloma sublateritium (Fr.) Karst. (69) Edible
"BRICK CAP"

Cap 2–10 cm broad, rounded to almost flat; brick red center, buff to cinnamon at margin; moist; margin inrolled at first. Flesh whitish brown bruising or maturing yellowish.

Gills attached, crowded; white to yellowish changing to gray to smoky purplish brown at maturity.

Stalk 5–10 cm long, 0.5–1.5 cm thick, equal; whitish above annulus darkening to brown below with flattened hairs.

Annulus thin, yellowish buff; partial veil fragments may also be attached to the cap margin, but they dry quickly and disappear.

Spores 6–7.5 × 3.5–4 μm, elliptical; spore print purple-gray.

Dense clusters on hardwood stumps and logs, late S., F.

Pholiota destruens (Brond.) Gill. (70) Not recommended

Cap 8–16 cm broad, rounded, often with a central knob; creamy maturing to light brown with scattered, large, soft whitish to buff scales; surface slimy when wet, often cracking across top; partial veil remnants hanging from the margin. Flesh firm; white.

Gills attached to notched, crowded; white to cinnamon at maturity.

Stalk 5–18 cm long, 1–3 cm thick, commonly enlarged at base; white above ring, densely scaly below and maturing darker tan to brown.

Annulus attached high on stalk near gills, loose, sometimes disappearing with age.

Spores 7–9.5 × 4–5.5 μm, elliptical, smooth; apical pore; spore print brown.

Single to densely clustered on ends of logs or on standing wood of cottonwood, aspen, and poplar, F.

71 *Pholiota squarrosa*

72 *Stropharia coronilla*

Pholiota squarrosa (Muller: Fr.) Kummer (71)   Edible?

Cap 3–12 cm broad, bell shaped to rounded broadening with maturity, yellowish to brownish yellow covered with tan to brown upturned scales in concentric rings; surface dry; margin wavy with partial veil remnants. Flesh firm; pale yellow.

Gills attached and sometimes notched, narrow, crowded; yellowish or tinged greenish brown finally dull rusty brown at maturity.

Stalk 4–12 cm long, 0.5–1.5 cm thick, equal; covered with tan to yellow-brown curved scales.

Annulus a fuzzy ring; partial veil membranous, whitish; rest of veil hanging from edge of cap, disappearing with age.

Spores 5–8 × 3.5–4.5 μm, elliptical, smooth; apical pore; spore print brown.

In dense clusters on hardwoods or conifers, F.

This species often has a faint onionlike odor and is eaten by some people when it is young; others have suffered gastrointestinal upsets.

Pholiota squarrosoides, a similar species, also occurs in clusters on wood. It has a scaly cap that is viscid between the scales and has slightly smaller spores.

Stropharia coronilla (Bull.: Fr.) Quél. (72)   Not recommended

Cap 2–5 cm broad, rounded to nearly flat at maturity; pale yellowish to yellowish brown; moist to viscid. Flesh thick, soft; white.

Gills attached, close; lilac becoming purple-black with white edges.

Stalk 2.5–4 cm long, 0.3–0.7 cm thick; white; minutely hairy above annulus, shiny longer hairs below.

Annulus white, soon purple-brown from spores.

Spores 7–9 × 4–5 μm, elliptical, smooth; apical pore; spore print purple-brown.

Several to abundant on lawns and in openings in woods, late S., F.

This species is suspected of being poisonous, thus **not** recommended.

Stropharia semiglobata (Batsch ex Fr.) Quél.

Cap 1–4 cm broad, rounded to conic flattening at maturity; bright straw yellow fading to dull yellow; slimy when wet. Flesh watery, yellowish.

Gills attached; olive-gray maturing deep purple-brown.

Stalk 2–10 cm long, 0.2–0.5 cm thick, often enlarged at the base; white to buff; hairy above annulus, slimy below.

Annulus delicate; whitish to pale yellow, often disappearing.

Spores 15–22 × 8.5–11 µm, elliptical, smooth; apical pore; spore print purple-brown.

Single to several, usually on horse or cattle dung, Sp., S., F.

Stropharia species generally should be **avoided** for eating. Many of them are poisonous.

Tricholomataceae

This is a large and diverse mushroom family, all genera of which have white to light lilac or yellowish spore prints and attached gills. Many of the Tricholomataceae occur on wood, and many mycologists use this as a nontechnical characteristic of this very diverse group.

Caps vary from laterally attached and fan shaped to rounded or flattened to depressed with a central stalk. Cap margins vary from even to inrolled, striate, wavy, or lobed. Flesh varies from white to shades of brown or gray.

Gills may be adnate, adnexed, or decurrent, close or crowded to distant, and white to brown or orange in color.

Stalks may be central to lateral or lacking. Texture and ornamentation also vary. *Armillariella* and *Cystoderma* have an annulus, but other genera lack this structure.

Spore prints are white or with light tints of lilac, buff, or yellow, and spore shape varies from nearly round through elliptical, cylindrical, or tapered at one end, smooth to roughened surfaces.

There are edible mushrooms in this group including *Pleurotus ostreatus* and *P. sapidus*. *Armillariella mellea* has long been considered edible by many persons but there are recent reports of gastrointestinal disorders associated with this mushroom. *Clitocybe dealbata, Omphalotus illudens,* and a number of species of *Collybia* are **toxic**.

KEY TO THE GENERA AND SPECIES IN THIS GUIDE
(Be aware of possible misidentification. Do NOT eat any specimen not positively identified.)

1a. Fruiting bodies develop on parasitized mushrooms, usually species of *Russula* and *Lactarius*. . . . .*Asterophora lycoperdoides*
 b. Fruiting bodies on wood, litter on ground, or from soil. 2

2a. Directly attached to wood, single, or in clusters 3
 b. On litter on ground or from soil 14

3a. Fruiting bodies usually in clusters 4
 b. Fruiting bodies single but sometimes grouped together. 9

4a. Fruiting bodies sessile or with lateral stalks; gills decurrent. . . . .
 . . . .*Pleurotus ostreatus*
 b. Fruiting bodies with a central stalk 5

5a. Fruiting bodies bell shaped when mature .6
 b. Fruiting bodies flat when mature 7

6a. Cap yellow to orange . . . . *Mycena leaiana*
 b. Caps reddish gray to reddish brown; blood red latex from stalks when broken . . . .*Mycena hematopus*

7a. Stalk velvety, brownish black . . . .*Flammulina velutipes*
 b. Stalk smooth, not brownish black 8

8a. Stalk and caps orange to orange-yellow; no partial veil. . . . .
 . . . .*Omphalotus illudens*
 b. Stalks and caps tan to shades of brown; partial veil cobwebby, becoming a poorly defined annulus . . . .*Armillariella mellea*

9a. Stalk lateral to almost lacking. 10
 b. Stalk central 12

10a. Cap with dark squamules, particularly at incurved margin . . . .
 . . . .*Lentinus tigrinus*
 b. Cap uniformly hairy, tough. 11

11a. Caps orange; sessile, or stalk eccentric. . . . .*Phyllotopsis nidulans*
 b. Caps violaceous to pinkish tan to reddish brown . . . . . .*Panus rudis*

12a. Cap tough; gill edges uneven . . . .*Lentinus tigrinis*
 b. Cap fleshy to firm; gill edges even. 13

13a. Cap white, mushrooms developing from knotholes on living trees, particularly elm and boxelder. . . . .*Pleurotus ulmarius*
 b. Cap with gray fibrils, on down logs or at base of stumps . . . .
 . . . .*Tricholomopsis platyphylla*

14a. Fruiting bodies attached to litter on ground or on basal bark of trees; caps thin; stalks slender, tough. 15
 b. Fruiting bodies developing through litter; caps fleshy 16

15a. Cap white; stalk black and wiry . . . .*Marasmius rotula*
 b. Cap orange; stalk brown and wiry . . . .*Marasmius siccus*

16a. Fruiting bodies in clusters or close groups. 17
 b. Fruiting bodies single or loosely scattered. 19

17a. Caps and stalks bright orange, usually near wood or developing from
     wood in soil . . . .*Omphalotus illudens*
  b. Caps and stalks not bright orange 18

18a. Stalks flattened; annulus lacking . . . .*Collybia dryophila*
  b. Stalks rounded; cobwebby annulus present, usually some evidence of
     it remaining . . . .*Armillariella mellea*

19a. Fruiting bodies developing near wood, such as down logs or
     stumps . . . .*Tricholomopsis platyphylla*
  b. Fruiting bodies widely scattered on ground 20

20a. Gills decurrent, at least to some extent .21
  b. Gills attached but not decurrent. 26

21a. Gills somewhat thick, waxy; cap cinnamon to reddish tan; or gills
     some shade of purple 22
  b. Gills narrow, not waxy; cap various colors 23

22a. Cap whitish to light gray; gills some shade of purple . . . .
    . . . .*Laccaria ochropurpurea*
  b. Cap cinnamon brown to reddish tan; gills colored like cap . . . .
    . . . .*Laccaria laccata*

23a. Cap deeply depressed, tan. . . . .*Clitocybe gibba*
  b. Cap rounded, flattened, or slightly depressed 24

24a. Cap bluish green, with aniselike odor, rounded to flattened . . . .
    . . . .*Clitocybe odora*
  b. Cap not bluish green, no distinctive odor 25

25a. Cap dull white, flattened to slightly depressed . . .*Clitocybe dealbata*
  b. Cap violet-gray; gills lilac. . . . .*Clitocybe nuda*

26a. Stalk with rooting base; cap viscid; in woods, single . . . .
    . . . .*Oudemansiella radicata*
  b. Stalk not rooting; cap dry. 27

27a. Fruiting in grassy areas, often in arcs or rings . . .*Marasmius oreades*
  b. Fruiting in woods . . . .*Tricholoma flavovirens*

73   A, B: Variations of *Armillariella mellea*

Armillariella mellea (Vahl.: Fr.) Karst. **(73A, B)** Edible
"HONEY MUSHROOM"

Cap 4–15 cm broad, convex to rounded sometimes knobbed; creamy tan or rusty brown to grayish brown with center darker; somewhat scaly with fibrillose squamules, viscid when wet; margin inrolled, finely striate. Flesh thin; white discoloring to brown with age.

Gills attached to slightly decurrent; white or dull cream staining to rusty brown with age.

Stalk 5–15 cm long, 0.5–2 cm thick, equal except enlarged slightly at base; dingy white with minute hairs above annulus, dingy becoming rusty below annulus.

Annulus cottony; white or buff to yellow; may break down with age.

Spores 7.5–9 × 5–6.5 μm, ovoid, smooth; spore print white.

Small to large clusters on living and dead hardwoods, particularly oak, or from wood buried in the soil, F.

A. *mellea* is very variable and the forms described above usually fruit on wood above ground in fall at time of leaf fall. However, an uncommon form fruits earlier from buried wood, producing large clusters of mushrooms with smooth, slightly sticky, honey-yellow to ocher-yellow caps on tapered stalks.

The mycelia of all forms of the "honey mushroom" are luminescent. A quiet walk on a dark September evening can be brightened by brushing aside oak leaves where the "honey mushroom" grows; glowing patches indicate the presence of the mycelia. Under certain conditions black, rootlike strands (rhizomorphs) develop from the mycelia and grow between the bark and wood of oak trees, hence the name "shoestring fungus." This form can be a damaging parasite.

Generally, A. *mellea* has been considered edible, but there have been reports of gastrointestinal upsets if the mushroom is eaten raw or incompletely cooked. We urge **caution** in eating this mushroom.

Armillaria mellea is an older name for this fungus.

74 *Asterophora lycoperdoides* on *Russula* sp.

Asterophora lycoperdoides Ditmar: S. F. Gray (74)

Cap 1–2 broad, subglobose to hemispherical; white becoming brownish; cottony becoming powdery. Flesh rather thick, moist; pale.

Gills often not developed, attached when present; whitish; sometimes forked.

Stalk 2–3 cm long, 0.3–0.8 cm thick, stout; whitish to brownish; stuffed becoming hollow, often curved.

Spores not developed when gills do not form; when present: 6 × 4 µm, elliptical, smooth; white. A completely different kind of spore, chlamydospores, develop from cells of cap: chlamydospores 12–18 µm in diameter, long spines give a star shape; brownish; accumulate in dry powdery mass on upper surface of cap.

Parasitic on *Russula* and *Lactarius* spp., common on *Russula nigricans*, late S., F.

Asterophora lycoperdoides was formerly referred to as *Nyctalis asterophora*. This parasitic mushroom may possibly be found on mushrooms of other genera but often the host mushroom is so disintegrated that it cannot be recognized.

A. parasitica is a similar parasite of other mushrooms, characterized by the production of large, smooth-walled chlamydospores.

Clitocybe dealbata (Fr.) Kummer ssp.   Poisonous
sudorifica (Peck) Bigelow

Cap 1.5–5 cm broad, convex flattening to shallowly depressed as it expands; white to whitish gray with vinaceous or pinkish tints when wet or maturing; surface dull, shiny or glabrous, at times subzonate; margin incurved, lifted when mature, irregular to lobed, or wavy but not striate. Flesh thin, brittle to pliant; white, buff when water soaked.

Gills adnate to short decurrent, crowded or close; whitish turning dingy pale buff.

Stalk 1–3 cm long, 2.5–5 mm thick at apex, equal or slightly enlarged at base; whitish maturing buff; surface with coating of fibrillose hairs; base white mycelioid.

Spores 4–4.5 × 2–3 µm, elliptical, smooth; spore print white to (rarely) cream.

Scattered or in groups, arcs, or rings in grassy lawns or pastures, S., F.

Clitocybe dealbata may fruit in or near fairy rings of *Marasmius oreades* and could easily be included with collections of that edible species.

At least two authors report severe gastric distress after eating this species, while others report it to be edible. This suggests the possibility that strains from different geographic locations may differ in toxin production. There is some disagreement about the interpretation of *C. dealbata;* however, all similar species are also considered to be **poisonous**.

75 *Clitocybe gibba*

Clitocybe gibba (Fr.) Kummer **(75)** Not recommended

Cap 3–9 cm broad, flat maturing to slightly depressed or depressed to funnel shaped; pinkish cinnamon to pinkish tan fading to whitish; moist, silky; margin inrolled at first, often irregular or lobed. Flesh thin; white.

Gills extending down stalk, crowded; white to cream to pale pinkish buff; sometimes forked.

Stalk 3–8 cm long, 0.5–1.5 cm thick, central; cap color or paler; dry, stuffed becoming hollow, smooth except densely hairy over the enlarged base.

Spores 5–10 × 3.5–5.5 μm, elliptical, smooth; spore print white.

Solitary to scattered under hardwoods and mixed woods, S., F.

This is a common, widespread mushroom that we hesitate to recommend for edibility because of the **danger** of confusing it with similar species that are poisonous.

Clitocybe nuda (Bull.: Fr.) Bigelow & A. H. Smith
Edible, choice

Cap 5–12 cm broad, rounded with inrolled edge flattening at maturity, sometimes with a small knob; violet-gray to lilac fading to pinkish brown at maturity; moist, smooth, hairless; margin uplifted, wavy. Flesh violaceous fading with age; odor faintly fragrant.

Gills notched at the stalk, crowded; cap color fading to flesh color at maturity.

Stalk 2–8 cm long, 0.5–1.5 cm thick, often enlarged and bulblike at base; cap color, darker at the top, nearly rusty brown at base; scattered white hairs on the lower part.

Spores 5.5–8 × 3.5–5 μm, elliptical, roughened; spore print pinkish buff.

Single to several in deep leaf litter under hardwoods or conifers, or in lawns, leaf piles, or compost piles, late S., F.

76 *Collybia dryophila*

Clitocybe odora (Bull.: Fr.) Kummer   Edible
"ANISE-SCENTED CLITOCYBE"

Cap 3–8 cm broad, rounded to nearly flat at maturity; bluish green to greenish gray to dull green to whitish. Flesh thin, soft; white to greenish white; odor and taste of anise.

Gills attached, decurrent, crowded; white to pale buff.

Stalk 2.5–7.5 cm long, 0.5–2 cm thick, often with enlarged base; whitish to buff to greenish; moist, downy above.

Spores 6–7.5 × 3–4 μm, elliptical, smooth; spore print pink to pinkish cream.

Scattered under hardwoods, S., F.

Fresh specimens of this species have a distinctive odor of anise.

Collybia dryophila
(Bull.: Fr.) Kummer **(76)**   Not recommended

Cap 3–5 cm broad, rounded flattening with age; rusty to tawny brown to reddish brown usually with a lighter band at margin; moist drying silky, smooth; margin often upturned, wavy. Flesh thin; white.

Gills attached to notched, narrow, crowded; white aging to buff.

Stalk 3–8 cm long, 0.25–0.5 cm thick, often swollen at base and surrounded by a mat of white mycelium; cap color or paler; slightly flattened.

Spores 4.5–6.5 × 3–3.5 μm, elliptical, smooth; spore print white.

Clustered to numerous in leaf litter under hardwoods, on wood or sawdust piles, Sp., S., F.

Toxin content and morphology of this species are quite varied. It is **not** recommended. Occasionally diseased fruiting bodies are found with varying amounts of distortion. Some have irregular, rounded, or lobed growth on the upper cap surface or gills; some are completely distorted.

77 *Flammulina velutipes*

78 *Laccaria laccata*

Flammulina velutipes
(Wm. Curtis: Fr.) Sing.   (77)

Edible

"WINTER MUSHROOM" or "VELVET STEM"

Cap 2–6 cm broad, rounded maturing nearly flat, yellow-orange to yellow-red to reddish brown, paler at edge; surface viscid, shining; margin upturned or wavy, ribbed. Flesh thick, thinning at edge; white to yellowish.

Gills notched, fairly well separated; white to yellowish; edges minutely hairy.

Stalk 2.5–8 cm long, 0.3–0.8 cm thick, often fused at base into clusters; yellowish tan at top, dense velvety brown to blackish brown below; tough.

Spores 7–9 × 3–4 μm, narrowly elliptical, smooth; spore print white.

Single or in clusters, often under the edge of bark, on down and live hardwoods, especially elm, Sp., S., F., W.

This is a species that can occur nearly anytime during the year, even during a winter thaw. This mushroom is also called *Collybia velutipes*.

Laccaria laccata (Fr.) Berk & Br.   (78)

Nonpoisonous

Cap 1.5–8 cm broad, rounded maturing flattened or upturned, often with a depressed center; reddish tan to cinnamon brown to pinkish; margin uneven, wavy. Flesh cap color.

Gills distant, irregular, extending slightly down the stalk; cap color or paler pink; waxy.

Stalk 5–10 cm long, 0.5–1 cm thick; cap color or dull flesh color, often with whitish to purplish base; covered with scattered loose scales in age, fibrous.

Spores 7.5–10 × 7–8.5 μm, broadly elliptical to nearly round; spines scattered; spore print white.

Single to scattered in moist woods, S., F.

One of the most common mushrooms in the woods in summer and one of the most confusing. It is variable and will be collected many times before it is recognized with certainty in the field. The widely spaced, pinkish, waxy gills are useful field characteristics; the spiny, nearly round, nonamyloid spores are a definitive microscopic characteristic.

L. amethystina is about the same size, but the entire mushroom is a beautiful deep violet color. *L. trullisata* is another somewhat similar species with a light brown cap and thick pink to purple gills. It develops in sand or very sandy soils.

79 *Laccaria ochropurpurea*

80 *Lentinus tigrinus*

Laccaria ochropurpurea (Berk.) Pk.   **(79)**   Edible

Cap 4–12 cm broad, rounded maturing to flat; whitish to light gray; dry. Flesh tough; dull white; taste unpleasant.

Gills attached to slightly decurrent, distant, thick; light to dark purple.

Stalk 5–12 cm long, 1–3 cm thick, more or less equal; cap color; often curved and twisted, dry.

Spores 8–10 μm, round, spiny; spore print light lilac.

Single to several under hardwoods, common under oak, late S., F.

This species tends to be larger than other Laccarias and can usually be distinguished by the size of the cap and the purple gills.

Lentinus tigrinus (Bull.: Fr.) Fr.   **(80)**   Not edible

Cap 2–6 cm broad, rounded, often depressed at maturity; buff to light cinnamon brown covered by brown-tipped to black squamules, which decrease in density toward the incurved margin; dry. Flesh tough; white.

Gills extending down the stalk, crowded; buff to pinkish tan; edges toothed.

Stalk 2–8 cm long, 0.4–1 cm thick, central to eccentric; buff with brownish scales becoming a dense hairy covering at the base, which is often rootlike extending into soil or rotting wood.

Spores 6.5–7 × 2.5–3.5 μm, oblong elliptical, smooth; spore print white.

Single or in dense clusters on hardwood stumps, logs, or buried wood, Sp., S., F.

Abortive fruiting bodies may occur alone or among normal ones. The gills remain covered by a veil, very few spores can escape. It was once thought that these abnormal fruiting bodies were a Gasteromycete, *Lentodium squamulosum*.

81 *Marasmius oreades*

82 *Marasmius rotula*

Marasmius oreades (Bolton: Fr.) Fr. **(81)**   Edible, choice
"FAIRY RING MUSHROOM"

Cap 2.5–5 cm broad, rounded maturing to bell shaped to flat often with a prominent blunt knob; cream to reddish tan fading when dry; dry, smooth; margin sometimes striate at maturity. Flesh thin; white or tan.

Gills notched, nearly free, well separated; paler than cap, creamy white to buff.

Stalk 2–8 cm long, 0.3–0.5 cm thick, equal; buff at top, darker brown at bottom with dense white hairs at base; tough.

Spores 7–11 × 4–7 μm, tapered at one end, smooth; spore print white.

Common, in groups or in fairy rings on lawns, pastures, and grasslands, Sp., S., F.

This mushroom has been eaten commonly and considered choice. It can also be dried for later use. There is some concern that it may take up some lawn pesticides, therefore the pesticide history of the area should be known and considered. Another rather serious concern is the possibility that *Clitocybe dealbata*, a poisonous species, may be found in the same area with *M. oreades*. Thus, *M. oreades* should be collected with care for the identification of each mushroom.

Marasmius rotula (Scop.: Fr.) **(82)**   Edibility unknown

Cap 0.3–1.2 cm broad, rounded with a central depression, distinctly folded; white; margin wavy, striate. Flesh very thin; white.

Gills not attached to stalk but to a collar encircling stalk, distant; white.

Stalk 1.2–5 cm long, 0.02–0.1 cm thick, equal; shiny black; tough, smooth, hollow.

Spores 7–10 × 3–5 μm, tapered at one end, smooth; spore print white.

Several to numerous on twigs and leaves on the ground in woods or on bark on lower trunk of trees, Sp., S., F.

Several other species of *Marasmius* with black wiry stalks and small white caps are not uncommon on leaf litter in the woods. Some have gills directly attached to the stalk.

Marasmius siccus (Schw.) Fr. **(83)**   Edibility unknown

Cap 1–2.5 cm broad, rounded to bell shaped, often with a depressed center; rusty orange-red to ochraceous red, darker in center; striate. Flesh very thin; white.

Gills notched to free, distant; white.

Stalk 3–8 cm long, 0.02–0.1 cm thick, equal; dark brown to brownish black at the base, paler at apex, often with white cottony mycelium at base; tough, smooth, dry.

Spores 13–18 × 3–4.5 μm, tapered at one end, smooth; spore print white.

Several to numerous on leaves, stems, or forest debris, S., early F.

This species is common, especially during humid weather.

83   *Marasmius siccus*

84   A, B: *Mycena hematopu*

Mycena hematopus
(Pers.: Fr.) Kummer   **(84A, B)**

Edibility unknown

"BLEEDING MYCENA"

Cap 1–4 cm broad, conic to bell shaped; reddish brown to pinkish brown, appearing frosted when young; margin striate, often ragged. Flesh thin; gray-red; latex dull red, exudes when broken.

Gills attached, close; whitish to grayish red, staining dark reddish brown.

Stalk 3–9 cm long, 0.1–0.2 cm thick, equal; pale cinnamon brown; minutely hairy, dry, brittle, hollow; dull red latex exudes when cut or broken near the base.

Spores 8–10 × 5–7 μm, elliptical, smooth, amyloid; spore print white.
Common, usually in clusters on much-decayed wood, Sp., S., F.

A

B

85 *Mycena leaiana*

86 *Omphalotus illudens*

Mycena leaiana (Berk.) Sacc. **(85)**   Nonpoisonous

Cap 1–4 cm broad, broadly rounded to bell shaped becoming expanded and convex, slight depression in the center; bright yellow-orange fading with age; viscid, smooth; margin striate or slightly so. Flesh soft, watery; white.

Gills attached, crowded; yellowish to salmon, bright orange-red at edges.

Stalk 3–7 cm long, 0.2–0.4 cm thick, equal or slightly enlarged at base; orange with fine hairs above, dense at base; viscid.

Spores 7–10 × 5–6 μm, elliptical, smooth, amyloid; spore print white.

Common, in dense clusters on hardwood logs and branches, S., F.

Omphalotus illudens (Schw.) Bresinski & Besi. **(86)**   Poisonous

"JACK-O'-LANTERN"

Cap 8–15 cm broad, rounded aging to flat, often with depressed center with a shallow knob; bright orange to orange-yellow; margin inrolled at first, even at maturity, streaked with appressed hairs. Flesh thin, firm; white to whitish orange.

Gills extending down stalk, crowded; yellow-orange to orange; sharp edged, luminescent in dark when fresh.

Stalk 8–12 cm long, 0.5–2 cm thick, tapering to narrow base; light orange; dry, longitudinally fibrous at maturity.

Spores 3–5 μm round, smooth; spore print creamy white.

Usually in large dense clusters at the base of hardwood stumps, or on buried wood or roots, especially oaks, late S., F.

Because this species contains a **violent** gastrointestinal **toxin**, great care must be used to **avoid** confusion with other similarly colored mushrooms or other fungi. *O. illudens* has been confused with *Cantharellus cibarius* even though *C. cibarius* fruits singly or scattered on the ground in the woods in midsummer. *O. illudens* has also been confused with *Laetiporus sulphureus*, "sulphur shelf." Both *L. sulphureus* and *O. illudens* occur in clusters on or very close to wood. They are easily distinguished, however, by the finely poroid to smooth lower surface of *L. sulphureus* in contrast to the narrow gills on the lower surface of *O. illudens*.

All three of these species produce fruiting bodies that are yellow to yellow-orange to red-orange in color, strikingly obvious. However, the bright color, the crowded decurrent gills, the densely clustered habit of growth, and the luminescent property of *O. illudens* characterize this species as distinctly different.

87 *Oudemansiella radicata*

88 *Panus rudis*

Oudemansiella radicata (Fr.) Sing. **(87)** Edible

Cap 3–10 cm broad, nearly flat, often wrinkled and darker at center, nearly white to gray-brown to dark brown; viscid or moist, smooth. Flesh thin; white.

Gills notched at stalk, broad, shiny, somewhat distant, of various lengths; white.

Stalk 5–25 cm long, 0.5–1.5 cm thick (above ground), slightly enlarged at soil line with a long, underground rootlike base; whitish at apex shading brownish to grayish at base; twisted-striate.

Spores 12–17 × 9–11 μm, ovoid, smooth; spore print white.

Single or scattered under hardwoods, occurring even in quite dry seasons, Sp., S., F.

This is one of the most common woods mushrooms throughout the growing season and is easily recognized by the following combination of characters: tapered underground base, small flat cap, long slender stalk, and widely spaced gills of different lengths. It is also known as *Collybia radicata* in many books.

Panus rudis Fr. **(88)**

Cap 1.5–7 cm broad, rounded to flattened, spathulate to fan shaped; violaceous to pinkish tan to reddish brown; dry, dense; stiff hairs covering the surface; margin inrolled and often irregularly lobed. Flesh thin, tough white.

Gills extending down the stalk, crowded; violaceous but soon appearing white to light yellow.

Stalk absent to short, when present, lateral to eccentric; cap color; hairy.

Spores 4.5–7 × 2.5–3 μm, short elliptical, smooth; spore print white.

Common, single, or clustered on hardwood stumps or fallen logs, S., F.

89 *Phyllotopsis nidulans*

A

90 *Pleurotus ostreatus.* **A:** On a standing tree trunk

Phyllotopsis nidulans (Pers.: Fr.) Sing. **(89)** Not edible

Cap 2–8 cm broad, fan shaped, rounded to flattened; light orange to orange-tan, surface densely hairy; margin inrolled to expanded at maturity. Flesh orange-buff to paler orange.

Gills adnate, crowded, narrow; orange-yellow or paler.

Stalk absent, sessile.

Spores 5–8 × 2–3 μm, elliptical, smooth; spore print light pinkish cinnamon.

Clustered on stumps and logs of hardwoods and conifers, S., early F.

This mushroom frequently has a strong, unpleasant odor, which makes it an unlikely candidate for eating. It is also known as *Claudopus nidulans*.

Pleurotus ostreatus (Jacq.: Fr.) Kummer **(90A, B)** Edible, choice

"OYSTER MUSHROOM"

Cap 2–20 cm broad, 1–10 cm wide, shell shaped to fan shaped, broadly rounded; white or pale tan to yellowish brown; smooth.

Gills attached and extending down the base, distant, broad, thick; white.

Stalk absent or much reduced, eccentric or lateral if present; white if present; often drier than the cap.

Spores 8–12 × 3–5 μm, elliptical, smooth; spore print white.

Several to clustered on branches, logs, or standing trunks of senescent or dead trees, usually hardwoods, especially elm, Sp., S., late F.

A delicious edible mushroom and easily recognized by the beginner.

Pleurotus sapidus (Schulze and Kalchbr.) Sacc. differs from *P. ostreatus* only by the lilac colored spore print. Late fall fruitings of *P. ostreatus* are often darker, gray to gray-brown.

There are several smaller to very small species on decaying logs with bracketlike, laterally attached caps or with sessile caps attached with the gill surface up. These white-spored mushrooms once were all interpreted as species of *Pleurotus*, but many are now considered members of other genera. They are all somewhat fleshy, usually occurring in large numbers if very small or in overlapping clusters if larger.

Resupinatus applicatus (Fr.) S. F. Gray, a common but very small sessile species, has caps 2–6 mm broad, grayish blue to grayish black; attached gill surface up; on underside of logs or larger pieces of decaying wood. They occur in large numbers and can dry and curl up only to expand when rewet. Even large fruitings are easily overlooked.

Pleurotus candidissimus Berk. & Curt. produces overlapping clusters of semicircular to shell-shaped, sessile, or laterally attached soft white mushrooms with widely spaced gills on decaying wood. It is probably the most common of the several small white species.

Panellus serotinus (Fr.) Kuehn, a fall species, has 2.5–6 × 2–3.5 cm wide, fan shaped, green to yellow-green viscid caps; pale orange gills; laterally attached in clusters on decaying logs.

Lentinellus ursinus (Fr.) Kuehn has larger, 2–10 cm long, and 2–5 cm wide, broadly attached, nearly flat, tan caps; few hairs on the upper outer portion with a dense covering of dark brown to cinnamon stiff hairs in the central area; gills coarsely toothed on lower surface, often with jagged edges; overlapping clusters.

Rhodotus palmatus (Bull.: Fr.) Maire has a 2–5 cm broad, brick red to pinkish, wrinkled cap with off-center short stalk; uniquely ridged surface with broad shallow pits; may occur singly or in clusters of several on down logs.

Pleurotus ulmarius (Bull.: Fr.) Kummer **(91)**   Edible
"ELM PLEUROTUS"

Cap 8–12 cm broad, rounded to flattened to somewhat depressed; white to buff; small scales on the surface; margin inrolled at first. Flesh firm; white.

Gills adnexed, moderately separated, broad; white to cream.

Stalk 5–10 cm long, 1.5-2.5 cm thick, nearly equal but sometimes expanded at the base; white to sometimes yellowish at maturity; dry whitish hairs at the top to nearly smooth at lower area.

Spores 5–8 × 4.5–7 μm, nearly round, smooth; spore print white to buff.

Single to clusters of two or three on hardwoods, particularly from branch scar sites on living elms and box elders, F.

This edible species has a very tough stalk. The flavor is somewhat like that of *Pleurotus ostreatus*, which we consider to be a flavorful mushroom. This late fall mushroom dries well, and can be seen on trees into the winter.

B

90 *Pleurotus ostreatus.* **B:** On a log

91 *Pleurotus ulmarius*

92 *Tricholoma flavovirens*

93 *Tricholomopsis platyphylla*

Tricholoma flavovirens (Fr.) Lundell   **(92)**

Cap 5–10 cm broad, convex becoming plane; pale to bright yellow; viscid, smooth to slightly scaly. Flesh firm; white or yellowish.
Gills adnexed, close to crowded; yellow.
Stalk 3–9 cm long, 1–2 cm thick, equal or slightly thickened at base; pale yellow to white; solid, smooth or slightly scaly.
Spores 6–7 × 4–4.5 μm, smooth, elliptical; spore print white.
Single to scattered on ground under mixed conifers and hardwoods, late S., F.

> *Tricholoma* is a large genus and the species are often difficult to definitely identify. They typically develop on the ground in late summer or fall and the mushrooms are usually rather large. Although some species are edible, some have an unpleasant flavor, and some are **poisonous.**
> *T. flavovirens* has been known as *T. equestre.* This is a distinctive species with its bright yellow color, but there are also other yellow species of *Tricholoma.*

Tricholomopsis platyphylla (Fr.) Sing.   **(93)**   Nonpoisonous

Cap 5–12 cm broad, flat to somewhat depressed at maturity; whitish gray or brown to blackish brown streaked with darker fibrils; margin at first inrolled, wavy with age. Flesh thin; white to gray; pliant.
Gills of various lengths, adnate to adnexed, distant, broad; off-white to white.
Stalk 6–12 cm long, 1–3 cm thick, equal or enlarging toward rounded base; white to grayish white as gills; smooth, hollow, or pithy with age.
Spores 7–9 × 4–6 μm, elliptical, smooth; spore print white.
Single to scattered on decaying logs of hardwoods or on well-decayed wood, appearing to be on soil, Sp., S., F.

> This species has also been called *Collybia platyphylla.*

Volvariaceae (Pluteaceae)

This family is represented by relatively few species in the Midwest, but some are spectacular and well known.

Caps are rounded to bell shaped, ranging in color from white to gray through shades of light brown and yellow to olive and even dark brown. The flesh is generally thin, white to off-white, soft texture, surface smooth to somewhat fibrillar (silky-hairy), and the margins are hairy-fringed or sometimes striate. Gills are free from the stalk, whitish when young maturing to pinkish, or even a brownish pink.

Stalks are fairly thick, vary in color but are often lighter shades of the cap color, and surface texture about as that of the caps. *Volvariella* typically has a prominent volva enclosing the base of the stalk.

The salmon to pink spores are elliptical and smooth. This feature along with the free gills are the most characteristic features of mature specimens. Spore prints likewise vary from light pink to salmon to pinkish brown in heavy deposits.

We are not aware of any poisonous species in this family, but neither have these been favorites for eating.

Pluteus cervinus (Schff.: Fr.) Kummer   **(94)**   Edible
"DEER MUSHROOM"

Cap 3–12 cm broad, conic to bell shaped or rounded to flattened at maturity; whitish, tan, or grayish brown, often with darker fibrils in streaks over the raised center; often slightly wrinkled, dull when dry. Flesh thin at the margins to thick at center; white.

Gills free, crowded; white becoming pink.

Stalk 5–15 cm long, 0.5–1.5 cm thick, enlarged at base; white tinted brown or dull brown with appressed hairs.

Spores 5.5–7 × 4–6 μm, elliptical, smooth; spore print salmon to pink.

Single to several on decaying hardwoods and conifers, sawdust, and buried wood, Sp., S., F.

Distinctive sterile cells, pleurocystidia, occur scattered among the basidia on the sides of the gills. They are large, 54–90 × 12–20 μm, fusiform ventricose with two to five tapered hornlike projections around the apex. This microscopic feature can be used to identify young or very old mushrooms of this species.

A number of other species of *Pluteus* occur on wood.

94 *Pluteus cervinus*

95 *Volvariella bombycina*

96 *Volvariella speciosa*

Volvariella bombycina (Schff.: Fr.) Sing. **(95)** Edible

Cap 5–20 cm broad, rounded to bell shaped; white to yellowish; silky fibrillose to shiny margin fringed with fine hairs. Flesh thick over cap center and thin at edges; soft; white.

Gills free; white becoming pink to rose at maturity; edges finely toothed.

Stalk 5–18 cm long, 1–2 cm thick, enlarged at base; white; silky, dry.

Volva deep enclosing base as a loose, cuplike bag; dull whitish discoloring yellowish.

Spores 6.5–10.5 × 4.5–6.5 μm, broadly elliptical, smooth; spore print salmon to pink.

Single to several on hardwood logs, in knotholes, or in crotches of living trees, late S., F.

Volvariella speciosa (Fr.) Sing. **(96)** Not recommended

Cap 5–15 cm broad, ovoid to globose expanding to convex or plane, more or less raised in the center; whitish to light tan, often darker in center; viscid, smooth, sometimes with pieces of universal veil adhering. Flesh soft, thinner toward margin breaking radially under rapid-drying field situations; white.

Gills free, crowded; white becoming deep flesh pink; edges somewhat uneven.

Stalk 9–20 cm long, 0.8–2 cm thick, sometimes enlarging towards base; white to cream; solid, smooth to slightly hairy.

Volva shallow; margin free from stalk; white to light gray.

Spores 11–20 × 7–12 μm, oval to ovoid, smooth; spore print deep flesh pink.

Single to small clusters in heavily manured or fertilized soils, especially in corn fields, in very wet springs.

Spectacular numbers of mushrooms may develop in a corn field one spring with no further indication of the fungus for years. This species may not actually be poisonous, but it is so similar to other species of *Volvariella* that have been reported to be poisonous that they could easily be confused.

Boletes

THIS group of fungi superficially resembles the gilled mushrooms, but instead of gills on the lower surface of the cap, the boletes have tubes, which are lined with a layer (hymenium) of basidia. Boletes have been traditionally considered to be members of a single family, Boletaceae, but modern interpretations divide them into several families. The arrangement of the open ends of the tubes, the pores (tube mouths), and the ornamentation of the stalk are field characters useful in determining the genera. Many species are mycorrhizal with various tree species, both softwoods and hardwoods.

Caps vary in shape from the typical rounded (convex) cap of *Boletus* to the sometimes fan-shaped, flattened cap of *Gyrodon*. Cap surfaces may be dry or viscid, hairy or floccose, with even to overhanging margins. Flesh colors vary from white or pale yellow to pinkish, and dramatic color changes from white to shades of red or blue upon cutting or bruising can occur in some species. Pore surfaces vary in color with maturity, often changing color as mature spores develop. Mature pore surface color is characteristic for a species and also may have color changes upon bruising or cutting.

Stalks tend to be fleshy like the caps and may be quite bulbous at the base. Ornamentation varies from none to reticulate networks to large scales (scabers). In several of the genera the partial veil becomes an annulus, which may be dry or viscid.

Spore print colors are characteristic of the genus.

There are some **poisonous** boletes, mostly species in the genus *Boletus*, which have red pore mouths when young and/or which bruise blue. Many boletes are edible and some choice. Some mushroomers recommend removing the tubes of edible boletes, particularly if they are old, because they may become slimy after cooking. Others recommend peeling the cuticle (outer layer of the cap) because of its consistency and texture. *Tylopilus felleus,* a striking fleshy bolete, unfortunately is so bitter that it is truly inedible.

KEY TO THE GENERA OF BOLETES IN THIS GUIDE

1a. Cap covered with coarse gray to blackish scales; spores globose to subglobose; spore print blackish brown to black . . . .
. . . . *Strobilomyces,* p. 165

 b. Cap lacking dark scales; spores elongate; spore deposit some other color. 2

2a. Spores small, oblong to short-elliptical. 3

 b. Spores long, elliptical to subfusiform 4

3a. Pores roundish; walls of even thickness; spore deposit pale yellow to yellowish ochraceous . . . .*Gyroporus,* p. 161

 b. Pores radially elongate; walls of varying thickness; spore deposit olive-brown. . . . .*Gyrodon (Boletinellus),* p. 161

4a. Spore deposit gray-brown, vinaceous or purple-brown; stalk not ornamented with dark points or squamules . . . .*Tylopilus,* p. 169

 b. Spore deposit colored otherwise; stalk ornamented or not 5

5a. Stalk roughened with points, dots, or squamules that darken at maturity, usually to very dark brown or black . . . .*Leccinum,* p. 163

 b. Stalk ornamented or smooth, ornamentation not as above 6

6a. Pileus viscid; spores elliptical . . . .*Suillus,* p. 165

 b. Pileus not viscid; spores elongate, inequilateral in most species . . . . .
. . . .*Boletus,* p. 159

KEY TO THE SPECIES IN THIS GUIDE

(Be aware of possible misidentification. Do NOT eat any specimen not positively identified.)

1a. Cap viscid (see also 11b) 2

 b. Cap not viscid 5

2a. Cap buff to yellow, orange-yellow, or brownish 3

 b. Cap red aging to buff with thick red hairs, dry . . . .*Suillus pictus*

3a. Cap buff, yellow to orange-yellow; lacking a ring 4

 b. Cap yellowish brown, often spotted; prominent ring. . . . .
. . . .*Suillus sphaerosporus*

4a. Cap bright yellow to buff; stalk narrow, curved; flesh bruising brown . . . .*Suillus americanus*
b. Cap orange-yellow, woolly at first to smooth at maturity, stalk thicker, not usually curved; flesh bruising blue to blue-green. . . . .*Suillus tomentosus*

5a. Stalk whitish to ivory to brown with network of ridges (reticulations). . . . .*Boletus edulis*
b. Stalks colored otherwise, ornamentation different or lacking. 6

6a. Stalk and cap covered by heavy gray to blackish, scaly hairs . . . .*Strobilomyces floccopus*
b. Stalk and cap ornamented otherwise or ornamentation lacking. . . .7

7a. Tube mouths elongate and uneven; tubes short . . . .*Gyrodon merulioides*
b. Tube mouths small, nearly round and even; tubes longer 8

8a. Stalks smooth, rusty rose to red except yellow at apex; cap deep red . . . .*Boletus bicolor*
b. Stalks not red, with or without ornamentation; cap not red. 9

9a. Stalk velvety, hairy. . . . .*Gyroporus castaneus*
b. Stalk with reticulations, projections, or dark dots 10

10a. Stalk with buff to light brown reticulations; tubes pink . . . .*Tylopilus felleus*
b. Stalk with dark tufted hairs; tubes not pink 11

11a. Cap margin with veil tissue clearly overhanging; cap dry, rusty red-orange to orange-tan. . . . .*Leccinum insigne*
b. Cap margin without overhanging veil tissue; cap dry or viscid, gray-brown to brown . . . .*Leccinum scabrum*

97 *Boletus bicolor*

98 *Boletus edulis*

Boletus bicolor Pk. **(97)** Not recommended

Cap 5–15 cm broad, rounded maturing to flat; deep red becoming rose red to pinkish red with age, yellowish at margin; dry. Flesh thick; pale yellow, slowly bruising blue.

Tubes 1 cm long, often depressed near the stalk; tube surface yellow aging to cream-yellow becoming olive as spores mature, slowly bruising blue; pores very small (1–2 mm in diameter at maturity), angular.

Stalk 5–10 cm long, 1–3 cm thick, even to tapered; yellow at apex, rusty rose lower portion; dry, smooth. Flesh yellow, bruising slowly blue.

Spores 8–12 × 3.5–5 μm, smooth, elliptical; spore print olive-brown.

Single to several or many under hardwoods (especially oaks) or in mixed forests on soil, late S., F.

Edibility of this species is **questioned** by some as certain individuals cannot tolerate it. It may be difficult to separate from other undesirable species on the basis of field characters.

Boletus edulis Bull.: Fr. var. *edulis* **(98)** Edible, choice
"STEINPILZ" or "KING BOLETE"

Cap 7–15 cm broad, convex; cream-brown to clay color to cinnamon buff to reddish brown; tacky becoming subviscid when wet. Flesh 2–4 cm thick, firm; white to reddish near cap surface, unchanging when bruised.

Tubes to 3 cm long, depressed near stalk; greenish yellow at maturity; stuffed when young; pores small (2–3 per mm), rounded.

Stalk 10–20 cm long, 3–4 thick at apex, 4–6 cm thick at base, clavate to more or less bulbous at base, massive; white to ivory to brown with network of fine lines (reticulations) especially at top.

Spores 13–17 × 4–6.5 μm, elongate-inequilateral, smooth; spore print olive-brown.

Solitary to scattered under hardwoods, S.

Smith considers *B. edulis* to be typically associated with conifers, but we find it commonly with no conifer association in oak-hickory forests. *B. edulis* is commonly parasitized by *Sepedonium chrysospermum,* and the entire bolete or portions of it may become a golden yellow from the masses of conidia.

99 *Gyrodon merulioides*

100 *Gyroporus castaneus*

Gyrodon merulioides (Schw.) Sing. **(99)** Edible

Cap 5–12 cm broad, nearly flat to slightly depressed; dull yellow-brown to dark reddish brown; dry becoming tacky when wet; margin incurved becoming uplifted and wavy. Flesh up to 1 cm thick; pale yellow, sometimes slowly bruising bluish green.

Tubes 3–5 mm deep, 2–3 mm wide, extending down stalk, strongly radiating; yellowish becoming dingy yellow bruising dark olive to reddish brown; pores elongate and uneven.

Stalk 2–5 cm long, 1–3 cm thick, eccentric to lateral; yellow-brown staining reddish on lower part at maturity or when injured.

Spores 7–11 × 5–7.5 μm, broadly elliptical to subglobose, smooth; spore print yellowish to brownish ochraceus.

Single to scattered, typically under ash trees, S., F.

The flavor is mild and not distinctive, and the appearance is not attractive. Often the stalks are short enough that the boletes are hidden in the grass under ash trees. This bolete is also known as *Boletinellus merulioides* or *Boletinus merulioides*.

Gyroporus castaneus (Bull.: Fr.) Quél. **(100)** Edible

Cap 2.5–8 cm broad, rounded maturing to nearly flat; chestnut brown to reddish brown to paler; velvety, dry. Flesh up to 1 cm thick white.

Tubes 3–7 mm deep, depressed around stalk; white aging yellowish bruising brownish; pores 1–3/mm.

Stalk 3–8 cm long, 0.6–1.5 cm thick, tapering upward to nearly equal; cap color but white below the tubes; dry, velvety, hairy, hollow.

Spores 7–11 × 4.5–6 μm, elliptical, smooth; spore print light yellow.

Single to several, very common under hardwoods, S., F.

Gyroporus cyanescens is larger with a cap 4.5–12 cm broad, pale yellowish to tan, and coarsely scaly to tomentose. All parts of the fruiting body, cap, the irregularly swollen stalk, and the tubes turn instantly blue when bruised. It occurs under hardwoods or in mixed woods.

101 *Leccinum insigne*

Leccinum insigne A. H. Sm., Thiers & Watling (101)

Edible

Cap 4–15 cm broad, rounded becoming broadly convex; rusty red-orange or orange-tan to dingy orange-brown; surface dry to subviscid, fibrillose to smooth; margin clearly exceeding the tube area (appendiculate). Flesh thick, firm becoming soft; white staining vinaceous gray when cut.

Tubes 1.5–2 cm long, depressed at stalk; pale yellowish green turning wood brown with a yellowish tan tint; pores small (2–4/mm), whitish staining dingy yellow or olive.

Stalk 8–12 cm long, 1–3 cm thick at the apex to 4 cm thick at base; surface whitish ornamented with brown scales (scabers) darkening to black at maturity, staining smoky gray when cut; underlying surface cottony, uneven.

Spores 13–16 × 4–5.5 μm, more or less fusiform, smooth; spore print dark yellow-brown.

Single to scattered with aspen in mixed woods, S.

The species we commonly find in the Midwest is a variant of *L. insigne*, which occurs under oaks. The cap generally is less orange and the stalk is narrower than the typical *L. insigne* of northern woods.

Leccinum scabrum (Bull.: Fr.) S. F. Gray

Edible

Cap 4–10 cm broad, rounded; grayish brown to yellowish brown; smooth becoming viscid; margin not extending past tube line. Flesh firm; white unchanging when cut or bruised.

Tubes 0.8–1.5 cm long, deeply depressed at stalk; white becoming wood brown when mature; pores small (2–3/mm), staining light yellow or unchanging when bruised.

Stalk 7–15 cm long, 0.7–1.5 cm thick, tapering to an enlarged base; whitish with dark brown to black scales (scabers) in a nearly networklike arrangement; dry.

Spores 15–19 × 5–7 μm, more or less fusiform, smooth; spore print olive-tan.

Single to several under birch, late S., F.

102 *Strobilomyces floccopus*

103 *Suillus americanus*

Strobilomyces floccopus (Vahl: Fr.) Karst. **(102)** Edible?
"OLD MAN OF THE WOODS"

Cap 4–15 cm broad, rounded, finally flat; whitish gray covered by grayish black; soft, coarse scales, dry; veil remnants hanging from the margin. Flesh white bruising pink or red turning to black.

Tubes 1.2–1.5 cm deep, white becoming gray to blackish staining like the cap flesh; pores large (1–2 mm), angular.

Stalk 5–12 cm long, 0.5–2 cm thick; cap color bruising black; shaggy, scaly, annular zone.

Partial veil soft, cottony; white to gray.

Spores 9–16 × 8–12 μm, nearly round, warty; spore print black.

Single to several under hardwoods, especially oaks or mixed woods, common in average seasons, present even in dry years, S., F.

If eaten, this bolete should have the cap **peeled** and the tubes **cut** away. It is not attractive, and few people seem to care to try it.

Suillus americanus (Pk.) Snell **(103)** Not recommended

Cap 3–10 cm broad, rounded to broadly convex, often with a small knob; bright yellow with buff to cinnamon patches of fibrils, sometimes streaked with reddish fibrils; margin with white to buff partial veil remnants, which disappear at full maturity. Flesh thin; yellow to orange-yellow bruising brown.

Tubes 4–6 mm deep; golden yellow to yellow-brown at maturity; pores large (1–2 mm wide), angular, extending down the stalk.

Stalk 3–9 cm long, 0.5–1 cm thick, equal, often curved; bright yellow, glandular dots especially on the upper half, darkening with age or upon handling; dry.

Partial veil cottony, thick, usually not forming a ring but hanging from cap margin.

Spores 8–11 × 3–4 μm, narrow, fusiform, smooth; spore print dull cinnamon brown.

Several to many under eastern white pine, late S., F.

104 *Suillus pictus*

105 *Suillus sphaerosporus*

Suillus pictus (Pk.) A. H. Sm. & Thiers (104) Edible

Cap 3–12 cm broad, conic, rounded to flattened at maturity; red becoming buff with age, abundant reddish hairs; dry becoming tacky when wet. Flesh firm; yellow bruising slowly reddish.

Tubes about 5 mm long; yellow to brownish at maturity; pores large (0.5–2 × 5 mm) angular, radially arranged and extending down stalk.

Stalk 4–10 cm long, 0.8–2.5 cm thick, enlarged downward, with dull red fibrils fading to gray with age; annulus floccose.

Partial veil dry, hairy; whitish.

Spores 8–10 × 3.5–5 μm, narrowly inequilateral, smooth; spore print olive-brown becoming clay color to tawny-olive on drying.

Single to scattered under white pine (*Pinus strobus*), S., F.

This mushroom is found in the Midwest where white pines have been used in windbreaks, yards, and other plantings, and in some native white pine stands.

Suillus sphaerosporus (Pk.) A. H. Sm. & Thiers (105) Not recommended

Cap 5–15 cm broad, rounded to nearly flat at maturity; dull yellowish brown; smooth, slimy, viscid; margin often retaining pieces of the partial veil. Flesh thick; white to yellowish staining brown.

Tubes 4–10 mm long, yellow staining brown; pores large (1/mm), angular, becoming decurrent near stalk.

Stalk 4–10 cm long, 1–3 cm thick; yellow staining brown; annulus gelatinous.

Partial veil large, often much torn, thick, tough; cap color or darker; outer layer gelatinizing.

Spores 6–9 × 6–8 μm, globose to subglobose; spore print olive-mustard when fresh, darkening with age.

Under hardwoods, particularly associated with oaks, S., F.

106   *Tylopilus felleus*

Suillus tomentosus (Kauf.) Snell, Singer & Dick.

Not recommended

Cap 5–15 cm broad, rounded to flattened at maturity; orange-yellow to buff with patches of gray-brown to reddish tomentum or squamules. Flesh thick; cream to yellow bruising blue or greenish blue.

Tubes 1–2 cm long, dingy brown; pores small (2/mm).

Stalk 4–10 cm long, 1–3 thick, equal to club shaped; cap color, interior yellow staining slightly blue-green when cut; dry, glandular, dotted.

Spores 7–12 × 3–4 μm, narrowly elliptical, smooth; spore print olive-brown drying to dull cinnamon.

Single to scattered under two-needle pines in windbreaks, yards, or plantations, late S., F.

Tylopilus felleus (Bull.: Fr.) Karst. **(106)**

Inedible

"BITTER BOLETE"

Cap 5–15 cm broad, broadly rounded; dingy cinnamon when young, buff to light brown with age; dry but somewhat slimy after rain, smooth. Flesh to 3 cm thick, soft; white unchanging to bruising brownish.

Tubes 1–2 cm long; pores about 1 mm in diameter at maturity, often depressed at the stalk; whitish becoming deep pink.

Stalk 4–12 cm long, 1.5–3 cm thick, enlarged to bulbous below; pallid to pale brown, darker below; covered with network of ridges at least in upper portion.

Spores 9–15 × 3–5 μm, narrowly elliptical, smooth; spore print rose to pink.

Single to scattered in woods, sometimes near conifer stumps and under hardwoods, S., F.

This bolete is not poisonous but is so **bitter** that it is inedible. Apparently some people do not taste this bitterness.

Aphyllophorales

A number of common fungi with tough fleshy to leathery to woody fruiting bodies are members of the order Aphyllophorales. At one time the fungi included here were grouped into families on the basis of the form of the fruiting body and of the areas on which the spore-producing cells were developed, that is, features that were readily visible. Modern classification of this large and very diverse group is based largely on microscopic characters and is difficult to apply in the field.

We have attempted to present these fungi in their modern groupings, but without indicating the technical reasons for doing so. We have used as much as possible the traditional common names for the groups: chantharelles, smooth fungi, pore fungi, coral fungi, etc. See the key to major groups, p. 31, for the initial identification of these groups.

Cantharelloid Aphyllophorales

This distinctive group varies in form from deeply depressed funnel-form fruiting bodies to normal mushroom shapes. The cap (pileus) may be deeply depressed with a layer of spore-producing cells on the lower surface, which can vary from a smooth to a blunt-ridged surface.

Cap colors vary from cream or buff through oranges and yellows to gray or shades of purple or brown. Cap surfaces are dry to moist and smooth to scaly, with margins entire to lobed, wavy, or incurved. Flesh colors vary from white or yellow to brown or nearly black. Ridges are broad, single to branched veinlike structures on the lower surface of the cap and tend to be clearly decurrent.

Stalks are usually equal, sometimes flattened or furrowed, and occasionally fused in small clusters.

Spore prints vary from white to pink or salmon-buff.

This group is known worldwide as one of the very best groups of edible fungi; most of them have excellent flavors. *Cantharellus cibarius*, the "golden chantarelle" of France, "kanterellas" in the Scandinavian countries, and "pfifferling" in Germany is found in many of the city markets in Europe when in season. It is highly prized throughout North America and is one of the best wild mushrooms in the Midwest. The "cinnabar" or "red chantarelle," *Cantharellus cinnabarinus*, that occurs in some numbers in northeastern Iowa and adjacent states is of excellent flavor.

KEY TO THE GENERA AND SPECIES IN THIS GUIDE

(Be aware of possible misidentification. Do NOT eat any specimen not positively identified.)

1a. Fruiting body yellow to orange 2
 b. Fruiting body purplish or gray or darker. . .*Craterellus cornucopoides*

2a. Ridges few, inconspicuous, or lacking . . . .*Cantharellus lateritius*
 b. Ridges well formed 3

3a. Cap color whitish yellow to orange-yellow. . . . .*Cantharellus cibarius*
 b. Cap color red-orange to cinnabar red . . . . .*Cantharellus cinnabarinus*

Cantharellus cibarius Fr.: Fr. (107)   Edible, choice
"GOLDEN CHANTARELLE"

Cap 3–10 cm broad, depressed with an inrolled margin when young, deeply depressed center with flat or upturned edges with age; light yellow to golden to orange-yellow; minutely hairy aging to smooth. Flesh thick; cream-yellow.

Ridges narrow, extending down stalk, blunt; often somewhat lighter than cap color; well separated, forked.

Stalk 3–8 cm long, 0.5–2 cm thick, tapered at base; near cap color; solid, smooth.

Spores 7.5–10.5 × 4–6 μm, elliptical, thin walled, smooth; spore print pale yellow to ochraceous.

Scattered to clustered in woods, under hardwoods or in mixed woods, especially in low areas and midslopes, S.

This species occurs fairly commonly during its season, is widely distributed, and is quite distinctive in appearance, and so it is not easily confused with other species; beginning collectors soon learn to recognize it with confidence. Recognize that the **poisonous** mushroom *Omphalotus illudens* is **similar** in shape and color, but it can be distinguished by its sharp-edged, knifelike gills. Also, *O. illudens* typically occurs in clusters very near or on down wood or stumps or at the base of trees.

107 *Cantharellus cibarius*

108 *Cantharellus cinnabarinus*

Cantharellus cinnabarinus Schw. **(108)** Edible, good
"RED CHANTARELLE"

Cap 1–4 cm broad, funnel shaped, cinnabar red to red-orange, dry, smooth; margin wavy, irregular. Flesh very thin; white.

Ridges narrow, decurrent on lower surface of cap, blunt; pinkish to nearly cap color; well separated, frequently forked.

Stalk 2–5 cm long, 0.5–1 cm thick, equal to slightly tapered at the base; cap color.

Spores 7–11 × 4–5.5 μm, thin walled, elliptical; spore print pinkish.

Single to several under hardwoods, S., early F.

This species is smaller and usually less common than C. *cibarius* but is similar in taste. Young fruiting bodies are quite red fading with age.

Cantharellus lateritius (Berk.) Singer Edible, choice

Cap 3–12 cm broad, convex to deeply funnel shaped; cream-yellow to orange-yellow; smooth to finely hairy; margin thin, inrolled at first to wavy when mature. Flesh thin; white to cap color.

Ridges low, inconspicuous, veinlike on the light cream-yellow to pinkish yellow lower cap surface.

Stalk 3–10 cm long, 0.5–2.5 cm thick, tapering to a narrowed base; cap color; dry, smooth, often hollow.

Spores 6–10 × 4.5–6.5 μm, elliptical, smooth; spore print pale pinkish orange (salmon) to pinkish yellow.

In clusters of varying size under hardwoods, especially oaks, S., early F.

This species has also been known as *Craterellus cantharellus*. The much-reduced ridges distinguish this species from C. *cibarius*, which may have a similar color and occur in similar habitats during the same season. Both species are edible.

Craterellus cornucopoides (L.:Fr.) Pers.   **(109)**   Edible

Cap 2–6 cm broad, deeply depressed funnel or trumpet shaped; purplish to ashy gray or brown to blackish outside, pale gray with dry stiff hairs on interior; margin widely flaring, wavy, and split. Flesh thin, brittle; dingy brown.

Lower cap surface slightly wrinkled to smooth, ash gray to blackish.

Stalk very short, sometimes absent; dingy brown; hollow when present.

Spores 8–11 × 5–6 μm, elliptical, smooth; spore print pale buff.

In clusters under hardwoods, especially oak and hickory, S., F.

Dark and rather unattractive, this fungus has several common names such as "horn of plenty," "trumpet of death," and "fairy's loving cup." It is reported to be edible, in contrast to its appearance, and is unlikely to be confused with other fungi. A similar appearing species, *Craterellus fallax,* has an ochraceous to pale salmon spore print and an almost smooth lower surface.

109   *Craterellus cornucopoides*

Clavaroid Aphyllophorales

The fungi in this group develop fleshy to tough fruiting bodies that are more or less branched stalks or fingerlike and are often known as coral fungi. The basal portion is sterile, but the upper clublike or branched portion is covered with a smooth layer of basidia.

The fruiting bodies are diverse in form and color, varying from pure white to tan, yellows, purples, and reds. Most are smooth to wrinkled, and branching may be diverse with tips varying from daintily pointed to broadly rounded or flattened. Flesh of the fruiting bodies is mostly white, but may vary in color.

Fruiting bodies develop principally among the leaves and litter on the forest floor, with a few common species inhabiting down logs. Species of *Thelephora* and *Tremellodendron* also occur on the forest floor, but have flattened branches.

Some coral fungi, including some species of *Ramaria,* are edible. However, there are several species that may cause gastrointestinal upsets; *Ramaria formosa* is considered to be **very poisonous**. This large, diverse group may be a confusing one for the beginner. Only a few very common species are discussed.

KEY TO THE GENERA AND SPECIES IN THIS GUIDE

(Be aware of possible misidentification. Do NOT eat any specimen not positively identified.)

1a. Fruiting body club shaped, unbranched 5
  b. Fruiting body branched 2

2a. Fruiting body with short, irregular, gray to purplish gray branches . . .
  . . . .*Clavulina cinerea*
  b. Fruiting body with longer, more slender branches 3

3a. Thin, slender branches with distinctive tips . . . .*Clavicorona pyxidata*
  b. Branches less slender, lacking distinctive tips 4

4a. Fruiting bodies on decaying logs, compactly branched with a number
  of similarly sized branches . . . .*Ramaria stricta*
  b. Fruiting bodies developing through litter on ground, branches of quite
  different diameters . . . .*Ramaria aurea*

5a. Clubs thin; white to golden yellow 6
  b. Clubs thicker than 1 cm in diameter, yellowish to ochraceous to light
  brown, often single . . . .*Clavariadelphus pistillaris*

6a. Fruiting bodies white browning at tip with age, usually clustered. . . .
. . . .*Clavaria vermicularis*
  b. Fruiting bodies golden yellow, in clusters . . . .*Clavaria pulchra*

Clavaria pulchra Pk. **(110)**   Edibility unknown

Fruiting bodies 1.5–7.5 cm high, 0.1–0.6 cm thick near the top, club shaped to nearly cylindrical, apex bluntly rounded; golden to egg-yolk yellow; sometimes smooth, sometimes grooved. Flesh tough, elastic.

Stalk not clearly distinct from apical spore-bearing area except for the lighter color.

Spores 4.5–7 μm, oblong-ovoid; spore print white.

Gregarious, clusters closely scattered on soil or litter in hardwood or mixed woods, S., F.

Two other groups of fungi form club-shaped fruiting bodies that could be confused with these unbranched corals. Some of the earth tongues, a group of ascomycetes, produce flattened to club-shaped, short-stalked fruiting bodies in damp or wet areas, often among mosses. There is an ascomycete genus, *Cordyceps*, that parasitizes insect larvae and pupae in the ground and produces club-shaped structures aboveground. Dig carefully to expose the base of any single or grouped club-shaped fungal structures developing from the soil to examine the basal portion for identification.

Clavaria vermicularis Fr. **(111)**

Fruiting body 3–10 cm high, 0.1–0.4 cm thick, club shaped or somewhat flattened, thickest at middle or above, tapering downward; white browning at tip with age. Flesh white.

Stalk short, more or less distinct; white.

Spores 4.5–7 × 2.5–3.6 μm, ovoid, smooth; spore print white.

Usually clustered in small groups, sometimes single, in mossy shaded lawns, and in woods, S., F.

This species is generally inconspicuous unless it develops in large numbers. There are several similar small, white, single to clustered species.

110 *Clavaria pulchra*

111 *Clavaria vermicularis*

112 *Clavariadelphus pistillaris*

113 *Clavicorona pyxidata*

Clavariadelphus pistillaris (L.: Fr.) Donk (112)

Not recommended

Fruiting body 4–20 cm high, club shaped tapering from broad apex to narrow base, often wrinkled or grooved with top rounded and often depressed; yellowish to ochraceous to brownish. Flesh firm; light ochraceous; taste bitter.

Stalk somewhat differentiated; lighter in color.

Spores 11–16 × 6–9 µm, elliptical, smooth, thin walled; spore print yellowish white.

Single to gregarious on the ground under hardwoods, late S., F.

If these distinctive club-shaped fruiting bodies are laid on a piece of colored paper, covered, and left for awhile, a spore print can be obtained. Spore prints can be made in this way from other coral fungus fruiting bodies.

A similar species, *C. ligula,* develops singly or in clusters in conifer duff under a number of conifer species. The clubs are usually smaller, 2–6 cm high.

Clavicorona pyxidata (Fr.) Doty (113)

Edible, tough

Fruiting body 5–12 cm high, 6–10 cm wide, highly branched; tan to pale pinkish tan; branches slender, equal, each tipped with spikes forming a crown or shallow cup. Flesh firm; white, somewhat waxy appearance.

Stalk whitish to pinkish brown at base.

Spores 4–5 × 2–3 µm, elliptical, smooth; spore print white.

Crowded or in clusters on dead wood or rotting logs, S., F.

This is a very common fungus on down logs, easily recognizable even when dried because of its distinctive "crown" of branches from upper levels of branching. It is reported to be mild flavored and edible when young, but it quickly becomes tough.

114 *Ramaria aurea*

115 *Ramaria stricta*

Clavulina cinerea (Bull.: Fr.) Schroet.   Edible

"GRAY CORAL"

Fruiting body 3–10 cm high, 3–10 cm wide, highly branched; smoky gray to dark gray to bluish gray at maturity; branches irregular with blunt or pointed tips, compact, often wrinkled. Flesh firm, somewhat brittle; gray to white.

Stalk short.

Spores 7–10 × 5.5–7 μm, broadly elliptical, smooth; spore print white.

Scattered to clustered on the ground, often in moss, in woods after rains, S., F.

Ramaria aurea (Fr.) Quél. **(114)**   Not recommended

Fruiting body 11–13 cm high, 4–8 cm thick or thicker, highly branched; pale orange to orange-buff with tips yellower. Flesh fairly brittle to brittle.

Stalk 3–5 cm long, up to 4 cm wide; lighter than fruiting body.

Spores 10–12 × 3.5–5.5 μm, roughened; spore print yellowish to pale tan.

Single or in clusters through leaf litter under hardwoods, late S., F.

Although *R. aurea* is common, it usually does not occur in large numbers.

Ramaria stricta (Fr.) Quél. **(115)**   Inedible

Fruiting body 4–12 cm high, 4–8 cm thick, highly branched with main branches arising from thick base; buff-pink to tan; branches erect, compact, tapering with tips pale yellow to buff-pink to pinkish tan. Flesh tough; taste bitter.

Stalk thick, branched above, broadened below where attached to wood; whitish.

Spores 7–10 × 4–5 μm, elliptical, minutely roughened; spore print cinnamon to buff.

Single to clustered on decaying hardwood logs, S., F.

This species is common and easily distinguished from the other common coral on decaying logs, *Clavicorona pyxidata*.

Poroid Aphyllophorales

The pore fungi comprise a large, diverse group characterized by shelflike "conks" (sometimes with a stalk) or flattened (resupinate) fruiting bodies on wood. The fruiting bodies are usually leathery to corky or woody, but a few are fleshy with a layer of open-ended tubes on the lower surface. The open tube ends are called pores.

The fruiting bodies are frequently fan shaped and broadly attached, but many have lateral or central stalks. The upper surface may be smooth or varnished or densely hairy, ranging in color from white or gray to bright orange to browns and black. They may be either annual or perennial. The color of the flesh above the pore layer (referred to as context with the poroid fungi) ranges from white or tan to brown.

The fruiting bodies of different species of these fungi usually develop on down wood, dead branches of living trees, and living trees. Some species develop in the heartwood of living trees causing decay of the heartwood and development of a hollow tree. Other species grow in the sapwood of trees.

Pore surfaces are characteristic for a species and may be variously colored; some may change upon bruising. Color changes of the pore surface and flesh upon treatment with various chemicals, such as KOH, are useful in identification. Pore size, traditionally indicated as the number of pores per millimeter (mm), ranges from minute to large, and pore shape varies from round to angular to elongate to lamellate or "mazelike," as in *Daedalea*.

Stalks, when present, may be fibrous or woody and often are the same color as the main portion of the fruiting body.

The spores vary in color from white to tan to brown.

While most pore fungi are too tough to eat, there are some edible species: *Grifola frondosa* ("hen of the woods"), *Laetiporus sulphureus* ("sulphur shelf"). They are excellent when young but become tough with age.

KEY TO THE GENERA AND SPECIES IN THIS GUIDE
(Be aware of misidentification. Do NOT eat any specimen not positively identified.)

1a. Tubes on lower surface of fruiting body separate; interior fleshy to tough, red-marbled like fresh meat. . . . .*Fistulina hepatica*
 b. Tubes on lower surface fused; interior not like above 2

2a. Fruiting bodies a number of fleshy overlapping brackets becoming tough with age 3
 b. Fruiting bodies tougher, corky to leathery to woody; clustered or single 4

3a. Fruiting bodies bright orange on upper surface, yellow on lower; colors fading with age .... *Laetiporus sulphureus*
 b. Fruiting bodies gray to smoky brown, typically with many small laterally stalked caps; at the base of oak stumps or trees .... *Grifola frondosa*

4a. Fruiting body with a shiny, dark red-brown upper surface .... *Ganoderma lucidum*
 b. Fruiting bodies dull, sometimes with a crusted or hairy upper surface 5

5a. Fruiting bodies with a central or lateral stalk 6
 b. Fruiting bodies sessile or nearly so, sometimes like overlapping shelves 10

6a. Pores averaging 3 or less than 3/mm; fruiting bodies developing in spring or early summer 7
 b. Pores averaging more than 3/mm; fruiting bodies usually developing in late summer or fall 9

7a. Fruiting bodies large (6–30 cm broad); stalk usually lateral, black at base; upper surface with brown scales .... *Polyporus squamosus*
 b. Fruiting bodies smaller; stalk central or lateral 8

8a. Stalk central; cap golden brown or darker; margin hairy .... *Polyporus arcularius*
 b. Stalk central or lateral; cap reddish yellow to brick red; margin smooth .... *Polyporus alveolaris*

9a. Fruiting body brownish black, at least in center; margin paler; stalk black; usually on down logs or stumps .... *Polyporus badius*
 b. Fruiting body tan; stalk black at base; usually on branches .... *Polyporus elegans*

10a. Fruiting body with a single layer of tubes (annual) 11
 b. Fruiting body with an indefinite number of tube layers, one developing each year (perennial) 18

11a. Pores round to angular but small (2–6/mm) or pore walls breaking into jagged toothlike structures 12
 b. Pores elongate radially, straight or curved or almost gill-like 17

12a. Fruiting body dark brown to blackish brown, large, thick; margin thick, exuding drops of fluid when young .... *Ischnoderma resinosum*
 b. Fruiting body different colors; smaller margin, thinner 13

13a. Fruiting body white; margin rounded projecting beyond pore surface;
on living or dead birch trees . . . .*Piptoporus betulinus*
  b. Fruiting body not white; margin ending at edge of pore layer . . . 14

14a. Fruiting body thin, tough; context white 15
  b. Fruiting body thicker; context colored 16

15a. Pore layer on young fruiting bodies white to lavender; pores becoming jagged and irregular with age, lavender color persisting at outer
lower edge for some time . . . .*Trichaptum biformis*
  b. Pore layer completely lacking lavender colors; upper surface of fruiting body with color zones . . . .*Coriolus versicolor*

16a. Fruiting body surface, context, and tube layer orange-red to orange
. . . .*Pycnoporus cinnabarinus*
  b. Fruiting body surface, context, and tube layer yellow-brown to rusty
brown . . . .*Polyporus gilvus*

17a. Thick gill-like walls delimiting very elongated modified tubes; fruiting
bodies sessile, tough brackets . . . .*Gloeophyllum sepiarium*
  b. Pores varying from round to elongate resulting in a mazelike lower
surface . . . .*Daedaleopsis confragosa*

18a. Fruiting body broadly semicircular; ridges on upper surface; lower
surface white when young bruising brown; pores very small . . . . . .
. . . .*Ganoderma applanatum*
  b. Fruiting bodies hoof shaped growing thicker with age but not much
larger in diameter; young pore surface not white. 19

19a. Fruiting body concentrically furrowed; interior reddish brown, on
various living hardwoods. . . . .*Phellinus igniarius*
  b. Fruiting body not concentrically furrowed; interior creamy tan to yellow-brown, usually on living ash trees . . . .*Fomes fraxinophilus*

Coriolus versicolor (L.: Fr.) Quél. **(116)**

"TURKEY TAIL"

Fruiting body 2–6 cm broad, 0.05–0.3 cm thick, leathery, broadly attached in overlapping clusters; upper surface variably colored with narrow concentric zones of white, cream, yellow, red, or bluish green to black, silky to velvety; thin black layer below colored zones; lower surface white to yellowish or light brown with age. Context thin (rarely more than 1 mm thick); white to tan.

Tubes up to 2 mm deep; pores averaging 3–5/mm, angular; white to yellowish.

Stalk absent, fruiting body attached laterally.

Spores 4–6 × 1.5–2 μm, cylindrical, smooth, hyaline; spore print white.

Dense clusters on dead hardwoods or on wound sites of live trees involved in sapwood decay, very common, S., F.

This species is also known as *Polyporus versicolor* or *Trametes versicolor*. It produces a white rot of dead hardwoods. *Coriolus hirsutus* is another very common species with thin, leathery, bracketlike fruiting bodies. These are nearly unicolored grayish to yellowish to brownish with darker marginal zones and a slightly thicker context (1–6 mm thick).

116 *Coriolus versicolor*

117 *Daedaleopsis confragosa*

118 *Fistulina hepatica*

Daedaleopsis confragosa (Bolt.: Fr.) Schroet. **(117)**

Fruiting body up to 15 cm broad, to 2 cm thick, sessile, leathery and pliable when young becoming rigid and firm; upper surface grayish to brownish, finely hairy to smooth; lower surface whitish to pale brown, often staining pinkish when injured when young. Context firm; pale buff to brown.

Tubes radially elongated; pores 1/mm; pore surface mazelike.

Stalk absent, fruiting body broadly attached.

Spores 7–9 × 2–2.5 μm.

Single to clustered, causing a white rot of dead hardwoods, rarely on conifers, annual, S., F.

This species is also known as *Daedalea confragosa*. The fruiting bodies are quite variable in form, sometimes difficult to interpret. Another rather common species formerly called *Daedalea unicolor*, now identified as *Cerrena unicolor*, has a mazelike pore surface. The fruiting bodies are thinner with a black line in the context, and the tubes often break into irregular flattened teeth with age. It also occurs on dead hardwoods causing a white rot. *Daedalea quercina*, especially common on oak, causes a brown heart rot and has a hard, more massive fruiting body up to 20 cm broad and 1–8 cm thick.

Fistulina hepatica Schaeff.: Fr. **(118)**   Edible, good
"BEEFSTEAK FUNGUS"

Fruiting body to 20 cm broad, fan shaped; upper surface blood red to liver color; moist, sticky. Context 3–7 cm thick, zoned white to reddish oozing a red juice; fleshy becoming fibrous and duller with age.

Tubes separate, closely packed; pores 4–6/mm, white to pale yellowish bruising dark, with a spongelike appearance.

Stalk short, usually lateral, sometimes a broad tapered area.

Spores 4–5.5 × 3–4 μm, ovoid, smooth, hyaline.

On stumps or at the base of living hardwoods, especially oaks, causing a brown rot, F.

This species is not common in the Midwest. The same common name is used for some false morels.

119 *Fomes fraxinophilus*

120 *Ganoderma applanatum*

Fomes fraxinophilus (Pk.) Sacc. **(119)**

Fruiting body 7 cm long, 9 cm broad, 7 cm thick, convex to somewhat hoof shaped, sessile; white turning grayish black to brown; margin thick, rounded, pale brown or white. Context woody; pale to wood color to yellow-brown.

Tubes 0.2–0.5 cm long; pores 2–3/mm, circular; pore surface white to brownish.

Stalk absent, fruiting body broadly attached.

Spores 6–9 × 5–6 μm, elliptical to ovoid, hyaline; spore mass white.

Usually single on living or dead ash trees, but also found on sycamore, oak, and elm, perennial, with a new tube layer developed each year.

This fungus causes an extensive white trunk rot of living trees, thus it may be a concern for property owners of diseased trees. On living standing trees, fruiting bodies often develop from areas where branches have naturally pruned. It is not uncommon to see several perennial brackets from such sites on a single tree.

Ganoderma applanatum (Pers.) Pat. **(120)**
"ARTIST'S CONK"

Fruiting body a shelflike conk up to 25–30+ cm broad, broadly fan shaped to semicircular, sessile; pale gray to gray-brown, dull; hard, smooth or concentrically ridged. Context dark rusty brown. Tubes in more than one layer after first year (perennial); tube layers separated by brown contextlike layer. Pores small, 4–6/mm; pore surface bright off-white in young brackets bruising brown when scratched; pore layer may become yellowish when older.

Stalk absent, fruiting body broadly attached.

Spores 6–11 × 6–8 μm, ovoid, truncate at one end, with minute spines; spore print brown.

On living deciduous trees, often from wounds; also common on down logs or hardwood stumps, very common on living aspen, seldom on conifers. Conks perennial with new annual tube layer developing in summer.

This fungus produces a white rot in standing trees and is probably a factor in the death of senescent trees of numerous species. The young white pore surface has been used as a surface on which to etch pictures and designs, thus the name "artist's fungus," or "artist's conk." This species is also known as *Fomes applanatus*.

121 *Ganoderma lucidum*

Ganoderma lucidum (W. Curt.: Fr.) Karst. **(121)**

Fruiting body to 15 cm or more broad to 3 cm thick at base, sessile to centrally or laterally stalked; upper surface dark red-brown with thin varnished crust at maturity. Context zonate; creamy white becoming dark purple-brown.

Tubes up to 1 cm deep; pores 4–5/mm, circular to angular; pore surface creamy white when young maturing to light buff, bruising dark purple-brown.

Stalk sometimes developed centrally or laterally, often absent.

Spores 9–12 × 5.5–8 μm, elliptical, truncate at apex, almost smooth; spore print white.

Single or clustered on living hardwoods near ground line or on dead stumps and tree trunks, common on oak; causes a white root and butt rot of living hardwoods, annual.

A

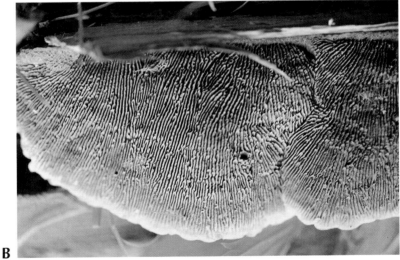

B

122 A, B: *Gloeophyllum sepiarium*

Gloeophyllum sepiarium (Fr.) Karst.   **(122A, B)**

Fruiting body 2–10 cm long, 1–7 cm wide, 0.3–0.8 cm thick, kidney shaped; upper surface rich rusty brown to red-brown or darker; hairy to somewhat smooth, zonate ridges; margin often white, yellow, or orange. Context yellowish to rusty brown.

Tubes to 0.7 deep, round to elongate to modified lamellate (gill-like) with thick walls; pores about 1–2/mm; pore surface golden brown to darker.

Stalk absent, fruiting body broadly attached.

Spores 6–10 × 2–4 μm, cylindric, hyaline; spore print white.

Single to several, often in clusters, on dead wood, usually on conifers but occasionally on hardwoods including birch, hawthorn, poplar, cherry, and willow.

This species causes a brown rot and often is the major fungus in the decay of dead conifers. Another species, *G. trabeum (Lenzites trabea)*, has a gray to pale, dingy cinnamon upper surface with a somewhat poroid lower surface, 2–4/mm. It usually is associated with a brown rot of hardwoods and with decay of both conifers and hardwoods used in building houses. It is also known as *Lenzites sepiaria*.

123 *Grifola frondosa*

124 *Ischnoderma resinosum*

Grifola frondosa
(Dickson: Fr.) S. F. Gray **(123)**

Edible, choice

"HEN OF THE WOODS"

Individual fruiting bodies 2–8 cm broad, 0.5–0.8 cm thick, fan shaped to petallike, sometimes fused; pale gray to gray-brown; margin thin, often wavy or curled, finely hairy to smooth; formed in a circular cluster as branches from a thick, cream-colored base, 10 cm thick or more in diameter, annual. Context about 2 mm thick in individual branches, fleshy becoming brittle; ivory-white.

Tube surface extending down stalk; pores small (1–3/mm), rounded; pore surface white aging yellowish.

Stalk white, lateral to eccentric, each single cap with a smaller stalk.

Spores 5–7 × 3.5–4.5 μm, ovoid, smooth; spore print white.

Clusters on soil near stumps of hardwoods, usually at base of oaks, F.

This species may require longer cooking than many mushrooms, but it is delicious.

Grifola frondosa causes a white rot and butt rot of living trees. *G. gigantea*, also known as *Meripilus giganteus*, looks like a larger and tougher *G. frondosa*. Individual fruiting branches, 10–30 cm broad, overlap to form a compound fruiting body 30–60 cm broad. Tubes and fruiting body flesh turn black when bruised or cut.

Ischnoderma resinosum (Schrad.: Fr.) Karst **(124)**

Fruiting body 7–25 cm broad, 0.8–4 cm thick; dark brown or blackish brown; densely hairy when young becoming smooth at maturity, sometimes shallowly furrowed, sessile; young margin thick and often exuding amber drops of fluid, thinner and dry with maturity. Context tough; straw color.

Tubes 1–10 mm deep; pores 4–6/mm; pore surface white to pallid maturing or bruising darker.

Stalk absent, fruiting body broadly attached.

Spores 4–7 × 1.5–2 μm, cylindric, smooth; spore print whitish.

Single to several, usually in overlapping clusters, on old logs and stumps of hardwoods and conifers, annual, late S., very common in F.

Exudate very conspicuous on edge of young fruiting bodies along the rather thick margin. *I. resinosum* causes a white rot of dead trees of numerous genera. It is also known as *Polyporus resinosus*.

125 *Laetiporus sulphureus*

Laetiporus sulphureus (Bull.: Fr.) Murr. **(125)**

Edible, caution

"SULPHUR SHELF"

Fruiting body single brackets 5–25 cm broad, up to 2.5 cm thick, usually forming clusters 30–60+ cm broad; upper surface bright orange to yellow when young fading to tangerine or pale brownish with age; lower surface sulphur yellow fading to pale tan; margin thinner with age, lobed, and wavy. Context up to 2 cm thick when young, tough to crumbly or chalky with age; white to yellow.

Tubes 1–4 mm long, shallow; pores 2–4/mm, roundish to angular; pore surface bright sulphur yellow to cream becoming pale tan.

Stalk lateral, very short or absent; whitish to yellow.

Spores 5–7 × 3.5–4.5 μm, broadly elliptical, smooth; spore print white.

In clusters on soil near stumps or on stumps, logs, or standing living hardwoods and conifers, very common on oak, annual, S. (occasionally early S.), F.

This common fungus, which causes heart rot of living trees, has also been known as *Polyporus sulphureus*. The shallow tubes develop late and young fruiting bodies appear smooth on the lower surface. The fruiting body fades to a pale whitish yellow with age.

Although edible for most people, gastrointestinal upsets have been reported for this species. The young edges of the fruiting bodies are most desirable for eating, the older portions soon become quite tough.

Phellinus igniarius (L.: Fr.) Quél.

Fruiting body 5–25 cm long, 3–15 cm wide, 2–12 cm thick, plane to convex hoof-shaped bracket; hard upper surface brown and finely tomentose when young, gray to blackish brown and smooth with age, concentrically furrowed, may be radially and irregularly cracked, sessile; margin thick, rounded, and pale. Context hard; reddish brown.

Tubes to 10 mm long, whitish gray maturing to gray-brown; new tube layers formed each year, older layers white stuffed; pores 4–5/mm; pore surface brown to grayish brown.

Stalk absent, fruiting body broadly attached.

Spores 5–7 × 4.5–5 μm, subglobose to globose smooth; setae rare to abundant; spore print white.

Single to several on hardwoods, perennial.

P. igniarius is found on trunks of a variety of living deciduous trees, common on birch and quaking aspen, and is quite variable. It causes a brown cubical rot of sapwood of living trees. It is also known as *Fomes igniarius*.

Phellinus everhartii is somewhat similar but occurs primarily on oak. *Phellinus pomaceus* is characterized by smaller brackets on *Prunus*.

126 *Piptoporus betulinus*

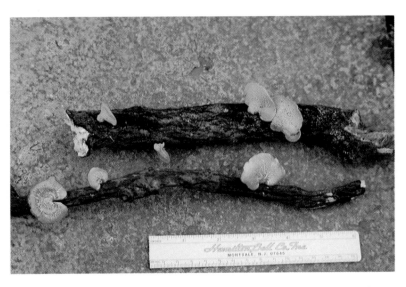

127 *Polyporus alveolaris*

Piptoporus betulinus (Bull.: Fr.) Karst. **(126)**

Fruiting body up to 20 cm broad, 2–5 cm thick; white to pale brown; smooth to slightly rough; margin inrolled, remaining as a rim around pore surface. Context tough when young maturing to corky, white.

Tubes 2–8 mm deep; tube walls becoming jagged with age; pores 3–4/ mm; pore surface white to cream becoming light brown with age.

Stalk lateral, short, or absent.

Spores 3.5–5 × 1–2 μm, straight or curved, smooth; spore print white.

Scattered on trunks of living or dead birch trees, common, annual, S., F.

This species causes a brown cubical rot of sapwood of dead birch. It is restricted to birch.

The short-stalked or sessile fruiting bodies are distinctive. *Fomes fomentarius* is a common perennial polypore on birch also, producing hoof-shaped fruiting bodies that become longer than broad with age. These two common fungi on birch are both easily identified.

Polyporus alveolaris (DC.: Fr.) Bord. & Sing. **(127)**

Fruiting body 1–8 cm broad, rounded to kidney or fan shaped; reddish buff or brick red weathering to cream or white, fibrillose, dry. Context thin, tough; white.

Tubes radially elongate in radial rows, walls breaking down with age; pores large (0.5–2 mm × 0.5–3 mm), angular to hexagonal; pore surface white to buff.

Stalk short, lateral or central; white to tan.

Spores 9–11 × 3–3.5 μm, narrowly elliptical, smooth; spore print white.

Single to several on branches and twigs of hardwoods, very common on shagbark hickory, S. and F.

Polyporus alveolaris is one of the group of pore fungi with large angular pores that form fruiting bodies in early summer. It causes a white rot of dead hardwoods. It is also known as *Favolus alveolaris*.

128 *Polyporus arcularius*

129 *Polyporus badius*

Polyporus arcularius Batsch: Fr. **(128)**

Fruiting body 1–8 cm broad, 1–4 mm thick, circular, depressed at center; golden or yellowish brown to dark brown with hairs; upper surface dry; margin hairy. Context thin, tough; white.

Tube surface extending down stalk; pores large (0.5–1 mm wide), angular, radially elongate; pore surface white to yellowish.

Stalk 2–6 cm long, 0.2 cm thick, central; yellowish brown to dark brown; dry, smooth.

Spores 7–12 × 2–3 μm, cylindric, smooth; spore print white.

Single or several in a group on hardwood sticks, logs, or stumps, early S.

This is one of a group of attractive species of poroid fungi with large angular pores that fruits in early summer. It causes a white rot of dead hardwoods.

Polyporus badius (Pers.: S. F. Gray) Schw. **(129)**

Fruiting body 4–20 cm broad, 1–8 mm thick, rounded to subcircular, convex becoming depressed; chestnut brown or reddish brown to nearly black at maturity; margin paler. Context thin, leathery; white or pallid.

Tubes to 3 mm deep; pores small (5–7/mm), circular to somewhat angular; pore surface white to tan to brownish.

Stalk 1–6 cm long, 0.5–1.5 cm thick, central or lateral; black at least on lower half; smooth.

Spores 6–8 × 3–4 μm, elliptical, smooth, hyaline; spore print white.

Single to several, very common on stumps and logs of hardwoods, late S., F.

It is not uncommon in fall to see large down logs completely covered with fruiting bodies of this attractive fungus. It causes a white rot of dead hardwoods and conifers. This species has been commonly called *Polyporus picipes.*

130 *Polyporus squamosus*

Polyporus elegans Bull.: Fr.

Fruiting body 1.5–7 cm broad, 0.2–0.5 mm thick, circular to oblong; pale to dull tan weathering to nearly white, nearly smooth, leathery drying rigid. Context white or pallid.

Tube surface extending down stalk; pores small (4–5/mm), angular; pore surface gray to tan.

Stalk 0.5–6 cm long, 0.2–0.6 cm thick, central to lateral; black at the base or lower half.

Spores 6–10 × 2.5–3.5 μm, cylindric, hyaline; spore print white.

Common, often on branches of hardwoods, late S., F.

This species causes a white rot of dead hardwoods. The fruiting bodies usually are smaller than those of *Polyporus badius*.

Polyporus squamosus Huds.: Fr. **(130)**　　　　Common

Cap 6–30 cm broad by 0.5–4 cm thick, roughly fan shaped; whitish to yellowish to dingy tan to buff; dry with large brown scales. Context thick, tough; white.

Tubes up to 7 mm deep, some deeply split with age; pores large (1–3 mm wide), angular; white to cream becoming straw color with age.

Stalk 1–4 cm thick, short, lateral to nearly central; white above, blackish below.

Spores 11–14–(20) × 4–5–(9) μm, elliptical; spore print white.

Single to several on hardwoods, from wounds on living trees, also on stumps and logs, Sp., early S.

Although *P. squamosus* is somewhat fleshy in appearance, it is tough even when young. It is the first species of the large angular pore group of pore fungi to produce fruiting bodies in the spring, often observed by morel collectors in May. This species causes a white heartrot of living and dead hardwoods.

131 *Pycnoporus cinnabarinus*

132 *Trichaptum biformis*

Pycnoporus cinnabarinus (Jacq.: Fr.) Karst. (131)

Fruiting body 2–12 cm long, 2–7 cm wide, 0.5–2 cm thick, sessile or broadly attached laterally; orange to cinnabar red fading at maturity; slightly tomentose weathering to glabrous. Context fibrous to soft-corky; red or orange-red.

Tubes less than 0.5 cm long; pores 2–4/mm, circular or angular; pore surface cinnabar red.

Stalk absent, fruiting body broadly attached.

Spores 4.5–6–(8) × 2–3 µm, cylindrical, often pointed at one end, smooth; spore print white.

Single to several on hardwoods, most common on oak and black cherry, occasionally on conifers, S., F.

This species is also known as *Polyporus cinnabarinus*. Its distinctive color makes it obvious and easy to recognize. This species causes a white rot of dead hardwoods.

Trichaptum biformis (Fr.) Ryv. (132)

Fruiting body 1–7 cm long, 1–7 cm wide, sessile often in overlapping groups; upper surface whitish to grayish to brownish, thin, tough, hairy, zonate at least at the margin; lower surface white to violaceous often fading with age but usually remaining violet at the outer edge. Context thin, usually less than 1 mm thick; white.

Tubes 1–3 mm long; pores 2–5/mm, round to angular when young, but soon breaking to become irregularly toothed; young pore surface white to violaceous fading with age.

Stalk absent, fruiting body broadly attached.

Spores 5–6 × 2–2.5 µm, cylindric to slightly curved; spore print white.

On dead hardwoods, common, S., F.

This species is one of a small group of thin, leathery, sessile pore fungi that develops in large groups of overlapping brackets on logs or stumps. They have color zones on the upper surface, weather slowly, and may persist for several years. Older weathered fruitings are difficult to interpret. The delicate violet color of the young fruiting bodies of *T. biformis* is unique. This species is also known as *Hirschioporus pargamaneus* and as *Polyporus pargamaneus*. This species causes a white pocket rot of dead hardwoods.

Schizophylloid Aphyllophorales

A small group of wood-inhabiting fungi, the schizophylloid Aphyllophorales produce small cuplike fruiting bodies that are attached at the base. A layer of basidia develops inside each cup. In some genera these small fruiting bodies are crowded together on a mat of fungal tissue. In others, such as *Schizophyllum*, the walls of the fruiting body enlarge, flatten, and elongate, exposing longitudinally split "gills." Typically tough and leathery, often with hairs on the outer surface, these fruiting bodies expand when moist and curl inward when dry.

Schizophyllum commune Fr. **(133)**

Fruiting body 1–3 cm broad, fan shaped, often attached broadly, sometimes in center of upper surface, densely hairy; whitish gray to gray; margin lobed and hairy. Flesh thin, tough.

Ridges gill-like, split, distant; white to gray; hairy.

Stalk short, lateral to absent.

Spores 3–4 × 1–1.5 μm, cylindrical; spore print white.

Several to overlapping in clusters on sticks, stumps, and logs of hardwoods, very common, often seen on senescent limbs of living trees, common on dead sections of living trees, Sp., S., F.

This fungus has been a favorite for experimental genetics research as it is readily cultured and produces fruiting bodies in the laboratory. The fruiting bodies curl on drying, becoming tough to almost brittle, then expand and flatten when moistened.

133 *Schizophyllum commune*

Smooth Aphyllophorales

This group of very diverse fungi typically inhabit or are associated with wood and are significant decay fungi of wood and wood products. They may develop on dead branches of living trees, but normally do not become established on healthy functioning wood. Development does not seem to be seasonal; thus they colonize wood whenever it is available. The fruiting bodies range in structure from thin sessile brackets to stalked fruiting bodies with flattened branches to smooth flat layers of a definite size or flat layers of various extent. Some of the flat fruiting bodies are a loosely interwoven, obscure hyphal tangle, and some have a more definite form.

Traditionally, these diverse fungi were grouped together because a flat uninterrupted layer of basidia that produces basidiospores develops on the surface of the fruiting structure. In the flat forms, the basidial layer develops on the surface away from the wood; in the brackets, the smooth basidial layer forms on the lower surface of the bracket. On the stalked branched forms, the smooth basidial layer covers the flattened branches. Microscopic features of the tissues of the fruiting bodies, of the cells of the basidial layer, and of the spores are diagnostic in modern interpretations of these fungi.

Stereum is one of the more conspicuous genera, developing thin bracketlike fruiting bodies in numbers on a down log or stick. Often the upper surface of the thin leathery brackets is zoned and very similar in appearance to the zonate, thin leathery brackets of the poroid Aphyllophorales. A look at the lower surface is sufficient to differentiate between the two groups for the smooth lower surface of the *Stereum* brackets is very different from the poroid pattern of the others.

Figures **134**, **135**, and **136** are some members of this group routinely encountered in deciduous woods and illustrate the diversity of the group. The tough fruiting bodies persist and are obvious all year, but are particularly evident in the fall when new fruitings occur.

134 *Stereum* sp.

135 Typical flat fruiting body of the smooth Aphyllophorales

136 *Stereum* sp.

Spiny, or Toothed, Aphyllophorales

Fruiting bodies in this group vary from branched, conk- or shelflike structures to those with caps having central or lateral stalks. The primary characteristic is toothlike spines that develop from the lower surface. The color varies from white through tan, brown, or orange. The flesh is soft to fibrous, tough, or brittle, and colors range from white to brown.

Stalks may be single, branched, or absent.

Spore prints are often white, but may be light tan to brown.

A number of species of this group are decay fungi of living trees or down wood. Those on down wood often form large, flat, thin fruiting bodies with conelike or irregularly jagged teeth.

The edible fungi in this group include *Hericium* spp. and *Hydnum repandum*.

KEY TO THE GENERA AND SPECIES IN THIS GUIDE

(Be aware of possible misidentification. Do NOT eat any specimen not positively identified.)

1a. Fruiting body caplike with stalk; spines from lower surface of cap . .3
  b. Fruiting body not caplike 2

2a. Fruiting body large, branched from a bulky base; spines clustered at the tips of the branches. . . . .*Hericium coralloides*
  b. Fruiting body unbranched, beardlike; spines from a lower surface of central fleshy mass. . . . .*Hericium erinaceus*

3a. Cap smooth, buff to cinnamon; teeth cream, even . . . .
  . . . .*Hydnum repandum*
  b. Cap surface with heavy brown scales, teeth brown. . . . .
  . . . .*Hydnellum scrobiculatum*

Hericium coralloides
(Scop.: Fr.) S. F. Gray   **(137)**

Edible, choice

Fruiting body 5–30 cm broad with many compact branches from a rooting base, branches tapering with clusters of delicate tapered spines hanging downward; white, aging dark cream; spines 5–40 mm in length. Flesh fibrous; white.

Stalk short, thick, branching.

Spores 4.5–5.5 × 5–6 μm, nearly round; spore print white.

Single to several on logs of hardwoods, late S., F.

Though the flavor may vary, this is usually considered to be an excellently edible fungus, either sauteed or cooked with meat and vegetables.

Hericium ramosum closely resembles *H. coralloides*, but differs by having shorter spines (3–10 mm long) on a more open branch system. It often has a more delicate appearance. Both occur on wood, develop in late summer or fall, and are edible. They are truly spectacular, especially when young and pure white.

137 *Hericium coralloides*

138 *Hericium erinaceus*

139 *Hydnellum scrobiculatum*

Hericium erinaceus (Bull.: Fr.) Pers.   **(138)**   Edible

Fruiting body large, 10–30 cm long, round fleshy mass with crowded slender spines hanging vertically from front edge and lower surface of fruiting body; white aging grayish or yellowish brown. Flesh firm; white. Stalk short to lacking.
Spores 4.5–6 × 4–5 μm, slightly roughened; spore print white.
From living hardwoods, especially oak, late S., F.

Hydnellum scrobiculatum (Fr.: Secr.) Karst.   **(139)**   Inedible

Fruiting body 3–7 cm broad, plane to funnel shaped; reddish brown zoned with lighter brown; solid; margin paler, uneven, sometimes lobed.
Teeth 1–2 mm long, brownish with gray tips; tooth layer extending down the stalk.
Stalk to 4 cm long, 1–3 cm thick; cap color, dull; sometimes fused with others to become irregular, mycelium at base interwoven with debris at soil surface.
Spores 4.5–6 × 4–4.5 μm, subglobose, warty, rough; spore print brown.
Single to clustered on ground under hardwoods or mixed woods, S., F.

There are several genera of fungi with centrally stipitate, tough, fibrous fruiting bodies with spines on the lower cap surface. *Auriscalpium vulgare* is widely distributed, typically fruiting on conifer litter or conifer cones. The fibrillar, small (1–4 cm broad) caps with brown spines on the lower surface and a central to laterally attached thin stalk are distinctive.
Fruiting bodies of *Hydnellum* and *Phellodon* are typically larger than *Auriscalpium* with thicker stalks, often with spines developing down the stalk. Species of *Phellodon* have white spores, those of *Hydnellum* have brown spores.

Hydnum repandum L.: Fr. **(140)**   Edible, choice

Cap 2–10 cm broad, rounded maturing to flat, depressed, often irregularly shaped; buff to tan to dull cinnamon; surface dry, smooth. Flesh white to cream, slowly bruising yellow.

Teeth 0.5–1 cm long, various lengths intermixed, extending slightly down the stem; white to cream darkening with age.

Stalk 2–10 cm long, 0.6–2 cm thick, equal to slightly enlarged at base; cap color; solid.

Spores 6.5–10 × 6.5–8 μm, nearly round, smooth; spore print white. Single to scattered under hardwoods, S., F.

This species is also known as *Dentinum repandum.*

140 *Hydnum repandum*

Gasteromycetes

THESE fungi develop basidia, the spore-producing cells, inside the fruiting body. The spores are not forcibly separated from the basidium, but the basidial cell collapses and the spores are dispersed in various ways that are characteristic of the different groups of Gasteromycetes. When spore color is described, it is spore color in mass, but it is observed directly in or on the spore-producing area of the fruiting body.

Fruiting bodies are diverse in this group, but are similar in each order of the Gasteromycetes. In the Lycoperdales, the puffballs and earthstars, the mature dry powdery spore mass is covered by a thin flexible cover that may develop an apical opening or break irregularly into large pieces. If the cover breaks and falls away, the spores are blown about by the wind or splashed about by rain. If the puffball has an apical opening, spores puff out through the opening when something pushes against the cover, such as drops of rain or a small animal scurrying through the woods.

In the Phallales, the stinkhorns, mature spores are in a foul-smelling slimy mass exposed on the fruiting structure. Insects that feed on the slimy mass carry the spores away on their bodies.

In the Nidulariales, the bird's nest fungi, the spores are produced inside the peridioles in the small cuplike fruiting bodies. The peridioles are flipped out of the cups by rain drops, but the spores are not released until the periodiole wall decays.

In the "hard puffball" members of the Sclerodermatales, the thick outer rind breaks irregularly exposing the dark powdery spore mass. The spores then can be blown about by the wind or splashed about by rain.

There are some species in the Lycoperdales that are considered excellent for cooking. Generally, their flavors are somewhat bland by themselves, but in soups, omelettes, and casseroles they seem to enhance other flavors. Especially recommended are *Calvatia gigantea*, *C. cyathiformis*, and *Lycoperdon pyriforme*. A word of **caution** should be added about the hard puffballs, species of *Scleroderma*. There are reports of illness by individuals who have eaten these fungi. The heavy rindlike cover and the early development of the purplish to purple-black areas in the interior of the fruiting body of a young *Scleroderma* are quite different from the thin cover and white interior shading into yellow to olive of a young true puffball.

KEY TO THE GENERA AND SPECIES IN THIS GUIDE
(Be aware of possible misidentification. Do NOT eat any specimen not positively identified.)

1a. Fruiting bodies with erect, spongy stalks covered with an olive-green to darker slimy spore mass at the apex (Phallales: Stinkhorns). 2
  b. Fruiting body sessile or stalked; stalk not spongy when present; spore mass dry 5

2a. Stalk pink to reddish with an indefinite apical region of slimy olive-green spore mass . . . .*Mutinus caninus*
  b. Stalk white with a distinctly limited olive-green to black slimy spore mass at apex. 3

3a. Stalk surrounded by an elongate, white netlike skirt . . . .
  . . . .*Dictyophora duplicata*
  b. Stalk lacking such a skirt 4

4a. Head portion covered by a smooth, olive-green to black slimy spore mass . . . .*Phallus ravenelii*
  b. Head portion with wrinkled ridge-pit network below the blackish olive slimy spore mass. . . . .*Phallus impudicus*

5a. Fruiting body resembling a tiny bird's nest with eggs (Nidulariales: Bird's Nest Fungi) 6
  b. Fruiting body round, pear shaped, or star shaped, rarely stalked. . .7

6a. "Nest" a short cylindrical cup with almost straight sides . . . .
  . . . .*Crucibulum laeve*
  b. "Nest" goblet shaped or vase shaped, smaller in diameter at base . .
  . . . .*Cyathus striatus*

7a. Outer layer of fruiting body splitting and recurving in starlike lobes; inner layer firm but flexible with a central opening (Earthstars). . . .8
  b. Outer layer of fruiting body not breaking into raylike segments, may break away at maturity or both layers may break in an indefinite manner at maturity 10

8a. Outer layer thick, almost woody, hygroscopic (rays bend out and
backward exposing inner layer when wet, curve in over it when
dry) . . . .*Astraeus hygrometricus*
b. Outer layer fleshy, tough when young, drying to become leathery, not
hygroscopic 9

9a. Inner layer surrounded by a ring of tissue broken from inner surface
of the fleshy outer starlike lobes. . . . .*Geastrum triplex*
b. Inner surface of outer starlike lobes unbroken, smooth . . . .
. . . .*Geastrum saccatum*

10a. Fruiting body with a definite, round to irregular opening (pore) in
central upper part of the intact inner layer. 11
b. Fruiting body not developing a pore; inner layer remaining or break-
ing irregularly at maturity 15

11a. Fruiting body with a tough, almost woody stalk . . . .
. . . .(Stalked Puffball) *Tulostoma brumale*
b. Fruiting body sometimes with a sterile, chambered basal portion;
lacking a definite woody stalk 12

12a. Lower part of fruiting body a sterile, chambered base, upper part
becoming a dry powdery spore mass (Puffballs) 13
b. Lacking a sterile basal portion or only a very small unchambered
area; dark powdery spore mass completely filling fruiting body
. . . .*Bovista plumbea*

13a. In clusters, often in large groups, on down wood; outer cover
smooth. . . . .*Lycoperdon pyriforme*
b. Single or loosely grouped, usually on ground, not wood 14

14a. Fruiting body with small pointed spines when young; mature spore
mass olive to olive-brown. . . . .*Lycoperdon perlatum*
b. Fruiting body with clusters of spines united at tips of spines; mature
spore mass purple-brown . . . .*Lycoperdon pulcherrimum*

15a. Cover of fruiting body thick, hard, persistent; capillitium lacking . . .
. 16
b. Cover of fruiting body thin; capillitium present or absent 18

16a. Outer cover very thick, eventually splitting irregularly and folding
back; fruiting body partially buried in soil . .*Scleroderma polyrhizon*
b. Outer cover thinner, breaking but not folding back; fruiting body at
soil surface 17

17a. Outer cover thin, flexible; fruiting body opening by an irregular slit . . . .*Scleroderma areolatum*
  b. Outer cover thick, distinctly scaly; fruiting body opening by irregular breaks . . . .*Scleroderma citrinum*

18a. Thin cover breaking, exposing rounded discrete inner units; these breaking to release dry, cinnamon spore mass . .*Pisolithus tinctorius*
  b. Thin cover breaking into irregular pieces at maturity, falling away from dry powdery spore mass and capillitium (larger Puffballs). . . . .
    19

19a. Fruiting body large (10–50 cm), roundish; outer cover breaking irregularly to expose dry powdery spore mass (completely fills fruiting body). . . . .*Calvatia gigantea*
  b. Fruiting body smaller, upper part larger in diameter; outer cover breaking from upper portion; sterile base . . . . .*Calvatia cyathiformis*

Astraeus hygrometricus (Pers.) Morgan   (141)

Fruiting body 1–4 cm broad when mature with expanded rays; outer cover tough, splitting into 7–15 pointed rays that bend and flatten outward when wet, curve inward over the continuous inner layer when dry; inner layer light gray to tan and opening by a slit or tear or irregular pore; interior white when young becoming a dry brown spore mass intermingled with dark threads (capillitium).

Spores 7–12 μm, nearly round, thick walled, covered with warts and spines.

Scattered in sandy soils and old fields, F.

The tough fruiting bodies will be persistent for several years.

Most of the "earthstars" are species of *Geastrum* and do not have hygroscopic rays that open and expand when moist and close when dry. However, one species, *Geastrum mammosum*, does have somewhat hygroscopic thicker and brittle rays and could be confused with *A. hygrometricus*. The spores are a distinctive differential character. Those of *G. mammosum* are about 4 μm in diameter, round, and much smaller than those of *Astraeus hygrometricus*.

141 *Astraeus hygrometricus*

142 *Bovista plumbea*

143 *Calvatia cyathiformis*

Bovista plumbea Pers. **(142)**

Fruiting body 2–3 cm broad, nearly globose; smooth, thin, white outer layer soon peeling off or breaking at maturity to expose a blue-gray to purplish brown inner cover with a narrow apical opening.

Spores 6–8 × 4–5 μm, oval with a long narrow pedicel (stalk) up to 13 μm long and loose threads (capillitium) several times branched with pointed ends; spore mass dark purplish brown, fills fruiting body.

Solitary to scattered in pastures or other grassy areas, S., F.

Young fruiting bodies are loosely attached to the soil by mycelium in the center of the lower surface. At maturity they are free and may be tumbled about by the wind.

Calvatia cyathiformis (Bosc) Morg. **(143)**   Edible, choice

Fruiting body 5–16 cm broad, 5–11 cm high, rounded; whitish to light tan; firm when young; upper portion maturing to a purple, dry powdery spore mass and sterile threads (capillitium); lower portion smaller, solid, firm, persistent; cover over upper spore mass cracks irregularly, breaks with age eventually falling away completely and exposing spore mass.

Sterile base, lower third remains long after top two-thirds, the capillitium and powdery spore mass, has disappeared.

Spores 3.5–7.7 μm, nearly round with minute spines; spore mass purple.

Scattered to numerous in grassy pastures, at edge of woods, late S., F.

A similar puffball, *C. craniiformis* **(144)**, also edible when young, has a bright yellow-green to yellow-brown spore mass at maturity. It is common on the ground in hardwood forests. The sterile bases of both species may persist for a year or longer.

145 *Calvatia gigantea*

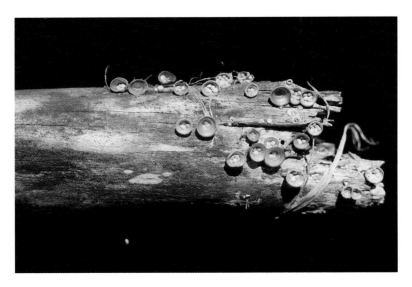

146 *Crucibulum laeve*

Calvatia gigantea (Batsch: Pers.) Lloyd   **(145)**

Edible, choice

"GIANT PUFFBALL"

Fruiting body 10–50 cm in diameter, round with a central basal cord-like rhizomorph often present; white or light gray to yellow then olive-tan at maturity; outer cover smooth, soft, leatherlike splitting randomly and breaking off at maturity completely exposing spore mass; interior white and firm when young becoming yellow to olive-brown powdery spore mass and threads (capillitium) at maturity.

Spores 4–5.5 × 3–5 μm, globose with minute markings; spore mass olive-yellow to olive-brown, fills fruiting body.

Single to several in arcs or irregular groups in mixed hardwood forests and meadow edges, late S., F.

Because of its size and appearance this is an easy fungus to recognize and when young is a choice edible. Some people do not like its texture.

Crucibulum laeve (Huds.: Rehlan) Kambly   **(146)**

"BIRD'S NEST FUNGUS"

Fruiting body 0.5–1.5 cm broad, 0.5–1 cm high, cup shaped, short cylindrical with sides essentially parallel; yellowish white to tan; externally velvety, initially with a tomentose membranous cover over the top of cup that soon disappears; contains 8–10 light yellowish tan or whitish peridioles (eggs), each 1–2 mm broad, disk shaped.

Spores 7–10 × 3–5 μm, elliptical, smooth.

On lignin-rich vegetable debris, stems, twigs, old nut shells, wood chips, S., F.

This species is also known as *C. vulgare*. The spores of this and other "bird's nest" fungi are released only when the thick wall of the peridiole weathers away.

147 *Cyathus striatus*

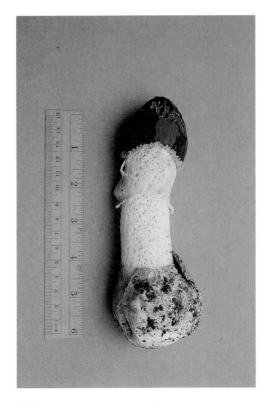

148 *Dictyophora duplicata*

Cyathus striatus (Huds.) Pers. **(147)**

"BIRD'S NEST FUNGUS"

Fruiting body 7–10 mm tall, 6–8 mm wide, vase shaped with slender base flaring in upper third at maturity; pale cinnamon brown to gray-brown; distinctly longitudinally ridged, hairy; white cover over top when young, breaking away exposing a striate interior wall and several drab to black peridioles (eggs), each attached to the wall by a cordlike structure; each fruiting body is attached by a mycelial pad.

Spores 18–20 × 8–10 μm, elliptical, thick walled, hyaline.

In clusters on bark and sticks or other woody debris, common on mulch under shrubs, Sp., S., F.

Cyathus stercoreus is common on dung and on manured soil. Fruiting bodies vary in size from 0.5–1.5 cm high and 4–8 mm wide with a shaggy outer cover on the smooth walls when young. The interior of open mature fruiting bodies is lead-gray to bluish black; the peridioles are black.

Dictyophora duplicata (Bosc.) E. Fischer **(148)**

Nonpoisonous, inedible

"NETTED STINKHORN"

Fruiting body 12–25 cm high, 2–6 cm broad at the base, tapered slightly upward to a head that is reticulate and olive-brown with a distinctive open depression at the apex; indusium (veil) white, netlike, attached below the slimy head, hangs 3–6 cm down stalk; young fruiting body (egg) 4–7 cm in diameter, globular to slightly flattened, white or flesh color, single or in groups.

Stalk cylindrical, honeycombed on the surface; white; spongy, hollow.

Volva white to brownish remains of outer egg cover.

Spores 3.5–4 × 1.5–2 μm, elliptical, smooth, hyaline.

Single to sparse clusters in humus under hardwoods and among shrubs, S., F.

The disagreeable odor characteristic of this "stinkhorn" increases with maturation and makes this otherwise attractive fungus an unlikely candidate for eating.

149 *Geastrum saccatum*

150 *Geastrum triplex*

Geastrum saccatum Fr. **(149)**

Inedible

"EARTHSTAR"

Fruiting body 0.6–2.5 cm broad, 8–15 mm high, ovate, pointed with a basal rhizomorph, two separate cover layers; outer layer golden tan to yellowish brown, splitting to form rays that recurve to the base, 2–5 cm in diameter when rays expand; sessile inner layer 0.5–2 cm wide with central opening (pore) surrounded by a paler area (mouth) bordered by a line, dull brown, flexible, dry, smooth; interior white, solid when young maturing to a brown mass of powdery spores and thick-walled encrusted threads (capillitium).

Spores 3.5–4.5 μm, round, warty; spore mass brown.

Scattered to clustered in rich humus around stumps in woods in late S., F.

Geastrum triplex Junghuhn **(150)**

Inedible

"EARTHSTAR"

Fruiting body 1–5 cm broad, with two separate cover layers; outer layer gray-brown to wood brown, fleshy, splitting to form 4–8 rays that recurve; fleshy inner part breaks forming a saucer- or bowllike structure surrounding the globose spore-producing area surrounded by the intact inner layer; inner layer with a central opening (pore) surrounded by a pale zone; interior initially solid, white maturing to a brown powdery mass of spores and thick-walled encrusted capillitium.

Spores 3.5–4.5 μm, round, minutely warted; spore mass brown.

Single to groups in rich humus in woods, often around well-rotted stumps, S., F.

151 *Lycoperdon perlatum*

152 *Lycoperdon pulcherrimum*

Lycoperdon perlatum Pers. **(151)** Edible, good
"PUFFBALL"

Fruiting body 1.5–6 cm broad, to 7 cm tall, obovoid to turbinate (upper portion larger in diameter); white maturing to creamy tan; surface covered with small spines that detach later leaving round, temporary scars that are lost as the fruiting body matures, opening by a central pore with age; interior white, upper portion becoming an olive to dark brown dry powdery mass of spores and modified hyphae (capillitium) at maturity; lower portion becoming a sterile base almost stemlike, chambered, and narrower than the upper area.

Spores 3.5–4.5 μm, round, minute spines; spore mass olive-brown.

Singly or in clusters on leaf litter and humus under hardwoods, common, late S., F.

All species of *Lycoperdon* are commonly called puffballs. When the flexible inner layer of a mature "puffball" is depressed by a large raindrop or a chipmunk scurrying along the ground or a person's finger, a puff of dry spores comes out of the central opening. One of the pleasant temptations of a walk in the woods in fall is the opportunity to help spread puffball spores!

All species of *Lycoperdon* are considered to be edible when young, when the interior is completely white.

Lycoperdon pulcherrimum Berk. & Curt. **(152)** Edibility unknown
"PUFFBALL"

Fruiting body 2–5 cm broad, 3–4.5 cm tall, pear shaped (pyriform) narrowing to a base about 1–2 cm; outer layer covered with a dense coat of white slender spines, which remain white even when dried and are often fused at their tips; mature spines fall from the purple-brown cover layer; upper two-thirds of young fruiting body white, spore-producing area (gleba) maturing into a purple-brown mass of spores intermingled with branched threads (capillitium); lower one-third develops into a base with distinct purple-brown chambers; apical pore with torn margins develops in the center of the top at maturity.

Spores 4–4.5 μm, globose, with stalk (pedicel) 10–13 μm; spore mass purple-brown.

Usually single to scattered under hardwoods, late S., F.

There are several other species of *Lycoperdon* with white spines that fall away as the "puffball" matures.

153 *Lycoperdon pyriforme*

154 *Mutinus caninus*

Lycoperdon pyriforme Pers. (153)   Edible, good
"PUFFBALL"

Fruiting body 1.5–3 cm high, 2–4 cm wide, pear shaped (pyriform) to subglobose, with conspicuous white rhizomorphs at base and in substrate; off-white when young, tan to brown at maturity; smooth, later covered with fine granules; round to slitlike apical pore formed when almost mature; upper two-thirds of interior matures into an olivaceous to olive-brown dry mass of spores intermingled with threads (capillitium); lower one-third a chambered base.

Spores 3–4 μm, round, smooth; spore mass olive-brown.

Single to dense clusters on rotted wood of any kind, common in late S., F.

Old weathered fruiting bodies are recognizable into the following summer. The only common species of *Lycoperdon* on wood, it often develops in large numbers completely covering rotting logs. White cords of mycelium (rhizomorphs) can be found throughout the wood.

Mutinus caninus (Pers.) Fr. (154)   Inedible
"STINKHORN"

Fruiting body 6–10 cm long, 0.8–1.1 cm thick, narrowly tapered top 2–3 cm of stalk with an olive-green, slimy spore mass; eggs 1–2 cm broad, 1–1.5 cm high, white subglobose to ovoid from which young fruiting body emerges and elongates.

Stalk honeycombed; pinkish to red, spongy.

Volva cuplike, remains of egg cover at base.

Spores 3.5–5 × 1.5–2 μm, cylindrical, thin walled; spore mass olive-green.

Single to several in soil and wood debris in late S., F.

This is a foul-smelling fungus! "Stinkhorn" is a most appropriate name.

155 *Phallus impudicus..* **A:** Immature; **B:** Mature

Phallus impudicus Pers. (155A, B)   Nonpoisonous
"STINKHORN"

Fruiting body develops from an egglike structure 3–6 cm in diameter **(A)**; pinkish to pink-purple, often with a cord from the center of the bottom; mature fruiting body 10–20 cm tall, 1.5–3.0 cm thick, distinctly reticulate covered by a blackish olive, slimy spore mass with an open depression at the top **(B)**.

Stalk 10–20 cm long, 1.5–3 cm thick; white to cream; spongy, honeycombed, hollow.

Volva cuplike, remains of tough egg cover around base of mature fruiting body.

Spores 3–4 × 1–2 μm, cylindrical; spore mass blackish olive.

Single or in clusters in wood debris or under hardwoods, S., F.

The rank, nauseating odor makes it possible to detect this fungus with ease. Carrion-inhabiting insects are attracted by the odor, feed on the spore mass, and incidentally carry the spores away on their bodies. The netlike (reticulate) appearance of the head region is more obvious after the dark slimy spore mass has been removed by the foraging insects or by rain. It is quite common in lawns, growing on dead tree roots or other woody debris that has been buried under a covering sod.

156 *Phallus ravenelli*

157 *Pisolithus tinctorus*

Phallus ravenelli Berk. & Curt.   **(156)**   Nonpoisonous
"RAVENEL'S STINKHORN"

Fruiting body develops from an egglike structure 3–5 cm broad, 2–3.5 cm high, often with a cord from the center of the bottom; pinkish or pinkish lilac; mature fruiting body 10–15 cm tall, 1.5–3 cm thick; head covered by an olive-green moist spore mass, 1.5–4 cm long, smooth to granular and with an open depression in the white apical disc.

Stalk 10–15 cm long, 0.5–3 cm thick; white to cream; spongy, honeycombed, hollow.

Volva cuplike, remains of tough outer egg cover around base of stalk. Spores 3–3.5 × 1.3–1.6 μm, cylindrical; spore mass olive-green.

Like *P. impudicus* this fungus has a foul odor, which makes it an unlikely candidate for eating even though it is nonpoisonous. This species seems to be less prevalent than *P. impudicus,* though both are found rather commonly.

Pisolithus tinctorius Pers.   **(157)**

Fruiting body 5–20 cm high, 5–15 cm thick, pear shaped to club shaped, tapered to a thick base with yellowish rooting fibers; dingy mustard yellow to golden brown to dingy brown; interior of numerous circular to oval chambers make up the upper half of the fruiting body maturing to a powdery mass from the top downward.

Spores 8–12 μm, round, with minute spines; spore mass cinnamon.

Single to several on packed soil along paths or roads in mixed woods or conifer plantings, S., F.

This heavy "puffball" (surprisingly dense) is foul smelling and stains hands and clothing an iodine color when rubbed. It is known to be mycorrhizal with some conifers and has been used commercially for this purpose.

158 *Scleroderma areolatum*

159 *Scleroderma citrinum*

Scleroderma areolatum Ehrenb. **(158)** Poisonous?

Fruiting body 1–5 cm broad, globose to irregular; firm outer cover (peridium) off-white or cream to light brown darkening to yellowish brown at maturity, dotted with distinct darker brown scales; fractures irregularly at maturity; entire interior of young fruiting bodies cream color developing purplish regions surrounded by irregular white lines, entire interior violaceous gray to darker grayish dusty spore mass at maturity.

Short stalklike enlargement at the center of the base of the fruiting body obvious or lacking; if present, ending in a group of short white to yellowish cords (rhizomorphs).

Spores 9–18 μm, round, variable, distinctly spiny; spore mass violaceous gray to more grayish.

Several to numerous on humus or wood debris under hardwoods, sometimes very common, typically irregularly scattered on ground in woods, late S., F.

All species of *Scleroderma* should be **avoided** because of reports of illness after eating some species. Species of *Scleroderma* are often called hard puffballs because of the firm outer covering that persists as a hard layer, opening in various ways, but often splitting irregularly at maturity, around the dry powdery, purple-brown to black spore mass. *S. areolatum* is also known as *Scleroderma lycoperdoides*.

Scleroderma citrinum Pers. **(159)** Poisonous?

"COMMON SCLERODERMA"

Fruiting body 3–12 cm broad, round to somewhat flattened at maturity; thick outer cover (peridium) yellow-brown covered with cracks, which accentuate the pattern of raised warts with somewhat darkened centers, sometimes breaking into irregular lobes when mature; entire interior becoming a fleshy, spore-producing area (gleba), lilac-gray to violet-gray with fine white lines throughout, at maturity a dry purplish-gray to black powdery spore mass.

Stalk merely a broadened portion of the outer cover in the center of the bottom, often with cords (rhizomorphs) attached.

Spores 8–13 μm, round, fine spines and ridges; spore mass brown to dark purplish gray to black.

Single to several or clustered on soil near the base of trees, around logs or stumps, or under hardwoods or conifers, late S., F.

This species has been known as *Scleroderma aurantium* Pers. and is one of the common hard "puffballs".

160   *Scleroderma polyrhizon*

161   *Tulostoma brumale*

Scleroderma polyrhizon
(Gmelin: Pers.) Pers. **(160)**

Poisonous?

Fruiting body 4–10 cm wide, round to somewhat depressed; outer cover (peridium) to 0.5 cm thick, rough, cream-yellow to yellowish tan to brownish to slate gray, at maturity splitting into irregular sections that eventually open and fold back; entire interior (gleba) at first dark purple interspersed with sterile white to yellowish cream lines, becoming a dark brown to black powdery spore mass at maturity.

Spores 6–12 µm, round, spiny, brown; spore mass dark brown to black.

Single or several on sandy soil, ditches, and under hardwoods, occasionally in lawns, F.

This species is also known as *Scleroderma geaster.*

Tulostoma brumale Pers. **(161)**
"STALKED PUFFBALL"

Fruiting body 1–1.5 cm broad, 1–2 cm high, subglobose with thin outer cover that sluffs away leaving a tan, smooth, flexible inner wall with a central opening (mouth); light reddish brown mass of spores and capillitium develop throughout interior.

Stalk 1–5 cm long, 0.2–0.4 cm thick with an enlarged base; pale tan; finely scaly to nearly smooth.

Spores 3–5 µm in diameter, globose, minutely warty, some with short stalks; spore mass reddish brown.

Capillitium 4–7 µm in diameter, threadlike, hyaline.

In sandy soil, F., persisting throughout the year.

Locally common in sandy areas of central Iowa, also in loess hills region of western Iowa.

The fruiting body of *T. simulans* is persistently covered with sand particles. That of *T. campestre* has persistent wartlike pieces of sand cover remaining for some time. The spores of these species are also different.

Jelly Fungi

T HIS artificial group of fungi, which includes species in the orders Tremellales, Auriculariales, and Dacrymycetales, is characterized by fruiting bodies that are gelatinous or jellylike to tough and are found on rotting wood or on the forest floor. The characteristic firm but gelatinous fruiting bodies of the jelly fungi are a result of the formation and accumulation of gelatinous compounds around the threadlike mycelia that interweave, forming the fruiting bodies. Technically these fungi are placed in different orders because their spore-producing structures differ in some morphological details, which are significant for the professional student of fungi but are of less concern for field recognition.

The jelly fungi produce fruiting bodies directly on the ground, on down wood, on the ground around the bases of living plants in the woods, or directly on rotting logs and leaf litter; a few species will grow on dead branches still attached in living trees. The fruiting bodies of some species have the ability to dry and become hard and brittle, then to absorb water and expand to normal size and texture when water is available; they can wet and dry for an indefinite number of times, potentially producing a crop of spores each time. Species of *Tremellodendron* produce rosettes of tough flattened branches, thus are quite different in appearance from most of the jelly fungi. They do, however, produce the kind of basidia characteristic of the order Tremellales.

This group is not known for its edibility, but "cloud ear," *Auricularia polytricha,* an Asiatic species, is eaten.

KEY TO THE GENERA AND SPECIES IN THIS GUIDE
(Be aware of possible misidentification. Do NOT eat any specimen not positively identified.)

1a. Fruiting body upright from ground in woods, branched 2

    b. Fruiting body developed on logs or woody branches, smooth or lobed 3

2a. Fruiting body tough and leathery; branches solid and typically flattened . . . . *Tremellodendron schweinitzii*
    b. Fruiting body tough gelatinous to cartilaginous; branches hollow and roundish . . . . *Tremella reticulata*

3a. Fruiting body ear shaped or lobed; drying to a hard brittle structure but reviving to be gelatinous and flexible 4
    b. Fruiting body firm gelatinous to tough, decaying with age . . . .
. . . . *Tremella foliacea*

4a. Fruiting body cuplike to ear shaped, yellow-brown to blackish brown; layer of basidia on lower surface only . . . . *Auricularia auricula*
    b. Fruiting bodies rounded to somewhat lobed, variously colored, broadly attached; basidial layer over upper surface 5

5a. Fruiting bodies light brown becoming darker to brownish black. . . . .
. . . . *Exidia glandulosa*
    b. Fruiting bodies bright orange to orange-red . . .*Dacrymyces palmatus*

Auricularia auricula (Hooker) Underwood **(162)** Edible

Fruiting body 5–15 cm broad, cuplike to ear-shaped lobes; yellow-brown when young becoming blackish brown and brittle when dry; centrally or laterally attached, tough gelatinous; upper surface with erect brown hairs; lower surface covered with basidia.

Spores 12–14 × 4–6 μm, oblong, curved, cylindrical, smooth; spore print white.

Singly or clustered on hardwoods, very common on shagbark hickory and elm, more common on hardwoods than on conifers, S., F.

This species has considerable variation. The older fruiting bodies tend to be lobed or ruffled, especially when clustered in tight groups on a down log. The ability to dry and become hard and brittle and to wet and return to a flexible, spore-producing condition increases the possibility for variable appearance.

162 *Auricularia auricula*

163 *Dacrymyces palmatus*

164 *Exidia glandulosa*

Dacrymyces palmatus (Schw.) Bres. **(163)**

Fruiting body to 3 cm in diameter, rounded to lobed, often wrinkled and convoluted, grouped in irregular clusters; bright orange to deep orange-red; at first flattened, tough gelatinous when young becoming softer, finally either soft and disappearing or more commonly drying as a hard orange to red mass; often in groups that expand to contact and then appear as a single fruiting body.

Spores 17–25 × 6–8 mm, cylindrical, curved, septate, developed on a Y-shaped divided basidium; spore print orange to yellow.

On coniferous wood, down logs, and branches, common, S., F.

A similar orange-yellow species, *Dacrymyces ellisii,* is common on down logs and branches of hardwoods. There also are yellow species of *Tremella* that with age typically become a soft watery mass and disintegrate. They produce spores on a septate basidium, a morphological feature that separates them into a different order, and have a white spore print. Any of these orange gelatinous jelly fungi have been called "witches' butter."

Exidia glandulosa Bull. ex Fr. **(164)**   Nonpoisonous

Fruiting bodies almost colorless initially, pustulate and separate becoming darker to brownish black then growing together and becoming broadly effused to 20 cm or more in length, drying black and brittle; spore-producing cells and wartlike papillae on outer surface.

Spores 10–16 × 4–5 μm, allantoid (curved); spore print white.

On decaying hardwoods, particularly oak and hickory, present all year.

This fungus is capable of repeated drying to a hard, horny fruiting body then reviving when moistened.

Another species, *Exidia repanda,* is also common on hardwoods. The discoid to lobed, centrally attached, smoky brown to cinnamon brown fruiting bodies are often clustered and fused. It becomes darker and brittle as it dries as does *E. glandulosa.*

165 *Tremella foliacea*

166 *Tremella reticulata*

Tremella foliacea Pers.: Fr.   **(165)**

Fruiting body 5–15 cm broad, leaflike with folds and branches flattened or tonguelike; tan to reddish brown or cinnamon; gelatinous to tough to nearly rubbery; interior lighter.

Spores 8–12 × 7–9 μm, elliptical to ovoid, smooth; spore print white.

Single to several on hardwoods, particularly oaks and hickory, S., F.

Several species of *Tremella* are very common on down logs throughout the summer and fall. One, *Tremella mesenterica*, produces fruiting bodies similar to the above except that they are golden yellow to orange-yellow. With age, the fruiting bodies of the various species of *Tremella* typically disintegrate into a soft, watery mass and disappear. *Dacrymyces* fruiting bodies are also tough and gelatinous.

Tremella reticulata (Berk.) Farlow   **(166)**

Fruiting body 6-8 cm tall, same wide, much-branched corallike structures to broadly flattened rosettes of hollow branches from a thickened, trunklike center, branches sometimes fused together; whitish becoming darker tan to yellowish tan with age; tough but gelatinous in appearance.

Spores 9–12 × 5–6 μm, elliptical to slightly curved, smooth; spore print white.

Single to several on ground under hardwoods, in wet periods, S., F.

Fruiting bodies may persist for several weeks before decomposing into a gelatinous dark mass. Older fruiting bodies parasitized by *Hypomyces rosellus* are covered by a rose-colored dry layer of compact mycelium.

Tremellodendron schweinitzii (Pk.) Atkinson   **(167)**

Fruiting body 2.5–12 cm tall, 5–15 cm wide, rosettes of leathery, upright flattened branches, often fused, arising from a thick, trunklike base; pallid to buff. Flesh cartilaginous, rubbery to tough.

Spores 7.5–10 × 4.5–6 μm, curved, cylindric, smooth; spore print white.

Single to clustered on the ground under hardwoods or in mixed woods, very common, S., F.

This species is tough enough to be very slow to decay. Old fruiting bodies may have some bright green areas or appear completely green because of the growth of green algae. Species of *Thelephora* also produce rosette fruiting bodies of flattened tough branches from a thickened base, but they are a purple-brown with gray tips when young. Both *Thelephora* species and *Tremellodendron* species differ from the true coral fungi by their tough, flattened branches and can be easily recognized even though they all typically develop on the ground in the woods.

167 *Tremellodendron schweinitzii*

Ascomycetes

A subgroup of Ascomycetes known as cup fungi have macroscopic fruiting bodies in which the asci are in an open layer on the upper surface of the cup (apothecium). However, there are other Ascomycetes that have tiny, round to flask-shaped fruiting bodies, smaller than 1 mm in diameter, with the asci inside. These tiny bodies may be produced on or in larger structures called stromata, which have distinctive shapes, colors, and textures, and are easily visible for field identification (see pp. 19–22).

KEY TO SOME COMMON ASCOMYCETES

1a. Fruiting bodies at least 1 mm in diameter, cup or saucer shaped, pitted spongy or convoluted, somewhat fleshy, with or without a stalk ....Cup fungi, p. 251
 b. Fruiting bodies less than 1 mm in diameter, round to flask shaped, sometimes buried in a stroma with only the opening of the fruiting bodies visible ....Other Ascomycetes, p. 281

Cup Fungi

KEY TO THE GENERA AND SPECIES OF CUP FUNGI IN THIS GUIDE
(Be aware of possible misidentification. Do NOT eat any specimen not positively identified.)

1a. Fruiting body with a distinct upper area and stalk, on the ground. .2
 b. Fruiting body sessile or without a distinctly different stalk, on ground or attached to wood....................................7

2a. Upper area flattened, spathulate, or fleshy with ridges, delimited pits, or convolutions. 3
　b. Upper area cup or saddle shaped . . . .(Saddle Fungi) *Helvella*

3a. Upper area with pits with sharp-edged ridges (spongelike) or convoluted with rounded lobes to almost smooth .4
　b. Upper area flattened, spathulate. . . . .*Spathularia flavida*

4a. Upper area completely attached to stalk or at least the terminal portion attached. 5
　b. Upper area attached only at end of stalk, hanging bell-like around stalk . . . .*Verpa*

5a. Upper area with sharp-edged ridges delimiting pits; completely or partially attached to stalk . .(Sponge Mushrooms, Morels) *Morchella*
　b. Upper area convoluted with smooth edges, or smooth and completely attached to stalk 6

6a. Upper area somewhat rounded to flattened, gelatinous, viscid, yellowish to greenish to olivaceous; stalk flattened. . . . . .*Leotia lubrica*
　b. Upper area convoluted with smooth edges, some shade of brown; stalk massive, often columnar . . . .(False Morels) *Gyromitra*

7a. Apothecium (fruiting body) on ground, sessile, free or attached to wood buried in soil or litter. 8
　b. Apothecium attached to logs or pieces of wood lying on surface of ground 14

8a. Apothecia sessile, on ground, free 9
　b. Apothecia developed at surface of soil or litter, attached to branches, pieces of wood or woody plant parts, on or in soil or litter 11

9a. Apothecia large, more than 1 cm in diameter; outer surface lacking stiff brown hairs 10
　b. Apothecia about 1 cm in diameter, deeply cup shaped; outer surface brownish covered with stiff brown hairs; inner area whitish . . . .
　. . . . .*Humaria hemisphaerica*

10a. Inner (upper) surface of apothecia orange. . . . .*Aleuria aurantia*
　b. Inner surface of apothecia typically tan to brown . . . .*Peziza*

11a. Apothecia large, attached to branches at least partially buried in leaf litter 12

b. Apothecia small, white to cream to pale yellow, growing on old acorns or hickory nut husks. . . . .*Helotium fructigenum*

12a. Apothecia brown 13

b. Apothecia with orange-red to red upper surface; lower surface whitish, sometimes with conspicuous white hairs . . . .*Sarcoscypha*

13a. Apothecia deeply urn shaped, dark brown; edges torn; flesh thin and tough . . . .*Urnula craterium*

b. Apothecia cup shaped with flat upper surface at maturity; edges smooth; flesh thick and gelatinous . . . .*Galiella rufa*

14a. Apothecia not green; flattened, scattered, or in crowded groups. .15

b. Apothecia green to olive or blue-green; with short stalks . . . .*Chlorosplenium aeruginescens*

15a. Apothecia brightly colored, smaller than 1 cm in diameter 16

b. Apothecia tan to brown, large, more than 2 cm in diameter. . . . .*Peziza*

16a. Apothecia bright yellow, usually in groups on rotting logs . . . .*Bisporella citrina*

b. Apothecia orange to orange-red, flat; brown stiff hairs around edge and on lower surface; often on soggy, well-rotted logs among mosses . . . .*Scutellinia scutellata*

168 *Aleuria aurantia*

169 *Bisporella citrina*

Aleuria aurantia (Pers.) Fuckel **(168)** Nonpoisonous
"ORANGE PEEL FUNGUS"

Apothecium 5–10 cm in diameter, often smaller, cup shaped and regular becoming broader, shallower, and contorted with age or from pressure, often split, sessile; exterior light orange to whitish; interior (hymenium) bright orange to reddish orange.

Asci 175–250 × 12–15 μm, cylindric, 8 spored; ascospores 18–22 × 9–10 μm, uniseriate; ends often overlapping, usually containing 2 large oil drops; wall reticulations regular and shallow to more ridged toward ends becoming pointed at each end, hyaline; paraphyses abruptly enlarged to apex to 7–8 μm, filled with orange granules.

Single to closely clustered on damp soils in woods and open areas, occasionally in lawns but more common on exposed clay soils or gravel paths, common, late S., F.

Bisporella citrina (Batsch: Fr.) Korf & Carpenter **(169)**

Apothecia shallow to flattened small cups, each 1–4 mm in diameter, centrally attached by small base; lemon yellow.

Asci 100–135 × 10 μm, cylindric, clavate, 8 spored; ascospores 9–14 × 3–5 μm, elliptical or fusiform, 1 celled, occasionally 1 septate, biseriate with oil drops at each end; paraphyses filiform, slightly enlarged at ends.

Clustered, often in large groups on decaying logs or pieces of wood, very common, S., F.

This species is also known as *Helotium citrinum* and is an obvious fungus on down logs in the fall, when the small cups occur in numbers sufficient to appear as bright yellow patches.

170 *Chlorosplenium aeruginescens*

171 *Galiella rufa*

Chlorosplenium aeruginescens (Nyl.) K. **(170)**

Apothecium 7 mm in diameter, cup shaped becoming flattened and expanded; blue-green.

Stalk short, 0.5–1.5 mm, central.

Asci to 70 × 5 μm, cylindric, clavate, 8 spored; ascospores 6–10 × 1.2–2 μm, fusiform, 1 celled, hyaline, smooth; paraphyses slender.

Scattered or in small clusters on down wood, mycelium staining wood blue-green, common on oak, S., F.

One often sees stained wood, particularly oak stumps, with no apothecia developed.

Galiella rufa (Schw.) Nannf. & Korf **(171)**

Apothecium 2–3 cm in diameter, shallow, cup shaped, sessile or stalked; margin at first incurved; exterior blackish brown covered with clusters of hairs, tough; interior gelatinous giving fresh apothecia a rubbery consistency; hymenium slightly concave becoming flattened, pale reddish brown; attached to sticks by a dense mass of black mycelium.

Stalk when present to 1 cm long, 4–5 mm in diameter.

Asci 275–300 × 12–14 μm, cylindric, 8 spored; ascospores 18–20 × 9–12 μm, elliptical with ends strongly narrowed, uniseriate, hyaline, granular; paraphyses threadlike, scarcely enlarged above.

Scattered or clustered on buried or partially buried wood, commonly on oak, in woods, S.

This species in also known as *Bulgaria rufa*.

172 *Gyromitra brunnea*

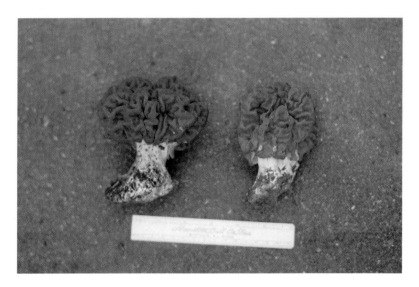

173 *Gyromitra caroliniana*

Gyromitra brunnea Underwood **(172)**   Poisonous?

Apothecium 5–12 cm broad and high, irregularly lobed to vaguely saddle shaped; reddish to dark brown above, whitish on under side.

Stalk 6–9 cm long, 2–5 cm wide, slightly enlarged at base; white; ribbed or fluted sometimes with anastomosing ridges, hollow or loosely stuffed.

Asci operculate, cylindric, 8 spored; ascospores 28–30 × 12–15 μm, elliptical, sculptured with fine warts or faint reticulations, hyaline, usually with 2–3 large oil drops at maturity; paraphyses numerous, slender, enlarged at apex.

This species is the common false morel in this area. After the fruiting body has reached mature size, the outer surface of the apothecium changes in shade of brown with age over the week to 2 weeks the asci are maturing. It has also been called *Gyromitra fastigiata* and *G. underwoodii*.

We do **not** recommend eating any species of *Gyromitra* at this time because of the possibility of the presence of the toxin monmethylhydrazine.

Gyromitra caroliniana (Bosc: Fr.) Fr. **(173)**   Not recommended

Apothecium 10–20 cm broad, 5–25 cm high, deeply wrinkled and convoluted; deep red-brown to brown.

Stalk 8–10 cm long, 3–5 cm wide, enlarged at base; white to cream; furrowed.

Asci operculate, cylindric, 8 spored; ascospores 25–30 × 12–14 μm, elliptical, usually with at least 2 large oil drops, hyaline, finely warted with 1 to several short apiculi; paraphyses numerous.

Single or loosely grouped on ground, in rich humus under hardwoods, late Sp.

These are impressively massive fruiting bodies, but not common. We do **not** recommend eating any species of *Gyromitra* at this time because of the possibility of the presence of the toxin monomethylhydrazine.

174 *Gyromitra infula*

175 *Helotium fructigenum*

Gyromitra infula Quél. (174)   Poisonous

Apothecium 3–10 cm broad, folded, saddle shaped, smooth to wrinkled but not convoluted, attached to stalk at several points; yellow-brown to reddish brown.

Stalk 5–10 cm long, 0.5–1.5 cm wide; tan to yellowish pink to darker; hollow, smooth to grooved or folded.

Asci operculate, cylindric, 8 spored; ascospores 20–23 × 7–10 μm, elliptical, with 2 large oil drops, hyaline, smooth; paraphyses numerous, slender, tips quite enlarged.

Single or scattered on ground under hardwoods or mixed woods, late S., F.

Contains the toxin monomethylhydrazine, which can cause serious **poisoning** as well as being **carcinogenic**. This species may well have a more northern distribution, but it has been collected in September in eastern Iowa.

Helotium fructigenum (Bull.) Karst. (175)   Inedible

Apothecium 1–4 mm in diameter, deeply cup shaped becoming flattened; white to cream to pale yellow.

Stalk 2–10 mm long, cream, slender.

Asci 80–100 × 7–9 μm, cylindric, clavate, 8 spored; ascospores 13–21 × 3–4 μm, oblong-fusiform or slightly inequilateral, uniseriate to biseriate, often with 2 larger and numerous smaller oil drops; paraphyses filiform, slightly enlarged above.

In groups on old acorns and hickory nut husks on ground in woods, often common, but inconspicuous, late S., F.

176 *Helvella crispa*

177 *Helvella elastica*

Helvella crispa Scop. ex. Fr. **(176)**

"WHITE SADDLE"

Apothecium about 5 cm wide, saddle shaped to reflexed, irregularly lobed and folded, attached at apex of stalk, white to pale cream.

Stalk 4–10 cm long, 2–3 cm wide, tapered upward; white to cream; deeply folded or fluted.

Asci operculate, cylindric, 8 spored; ascospores 18–20 × 9–13 μm, elliptical containing 1 large oil drop, hyaline, smooth; paraphyses numerous, cylindrical, enlarged above.

Solitary to clustered on ground under hardwoods or conifers, common mid-S., F.

Though *H. crispa* is common, it is usually not found in large numbers at one time.

Another species, *Helvella acetabulum*, also has a fluted stalk but the stalk is shorter and the apothecium is more cup shaped. It is common during spring in some years, and is often parasitized, entirely covered with the parasite so that the fruiting bodies are a dusty rose-pink.

Helvella elastica Bull.: Fr. **(177)**

"SADDLE FUNGUS"

Apothecium 0.5–5 cm long, 3 cm wide, saddle shaped to irregularly 2–3 lobed, centrally attached to stalk; whitish to medium gray to gray-tan, drying darker.

Stalk 5–10 cm long, 3–10 mm wide, cylindrical; white to yellowish; smooth.

Asci operculate, cylindric, 8 spored; ascospores 18–20 × 10–13 μm, elliptical, containing 1 large oil drop, hyaline, smooth at maturity; paraphyses numerous, clavate to occasionally branched.

Single or clustered on ground in wooded areas, common but usually not fruiting in large numbers at one time, S., F.

Several smaller species of *Helvella* with darker apothecia are less common.

178 *Humaria hemisphaerica*

179 *Leotia lubrica*

Humaria hemisphaerica
(Wiggers: Fr.) Fuckel (178)

Apothecium reaching 2–3 cm in diameter, about 1–1.5 cm deep, at first globose gradually expanding to become cuplike, hemispheric, sessile; inner surface (hymenium) whitish; outer surface brownish with stiff hairs that also rim margin.

Hairs 400–500 × 15–20 μm, enlarged at base tapering to sharp apex, septate, thick walled; dark brown.

Asci 325–350 × 15–18 μm, cylindric; ascospores 25–27 × 12–15 μm, elliptical, uniseriate, usually containing 2 large oil drops, hyaline to subhyaline, wall with minute roughenings; paraphyses slender, enlarged at apex.

Common on soil in woods, more rarely on rotten wood, S., F.

Leotia lubrica Pers. (179)　　　　　　　　　Inedible
"JELLY BABIES"

Apothecium 10–12 mm broad, somewhat rounded to caplike; olivaceous; somewhat viscid; margin lobed, overhanging.

Stalk to 6 cm long, 8–10 mm thick, flattened; ochraceous.

Asci about 150 × 10–12 μm, inoperculate, 8 spored; ascospores 20–25 × 5–6 μm, slightly fusiform with rounded ends, often slightly curved, becoming 5–6 septate, hyaline; paraphyses slender, branching with clavate tips, hyaline.

Usually in clusters of many stalked apothecia on soil among mosses in woods, often in moist shaded areas, not uncommon, S.

180 *Morchella angusticeps*

181 *Morchella crassipes*

Morchella angusticeps Peck **(180)**

"BLACK MOREL"

Apothecium 2–9 cm long, 1.5–5 cm wide, caplike, narrowly conic when young becoming broader with age, spongy; pits elongate, brownish gray to grayish brown with ridges dark brown to black.

Stalk 1.5–6 cm long, 0.5–3 cm wide; whitish to creamy to pinkish tan; scurfy, hollow.

Asci 200–300 × 16–22 μm, operculate, 8 spored; ascospores 21–24 × 12–14 μm, elliptical, 1 celled; spore print cream.

Scattered to clustered on ground under conifers or hardwoods, S.

This species is one of the "black" morels; another species occurs in western United States. Both have been reported to be edible and choice, but there are some reports of **mild poisonings** from the black morels.

Morchella crassipes (Ventenat: Fr.) Pers. **(181)** Edible

"THICK-FOOTED MOREL"

Apothecium 5–18 cm long, 4–8 cm wide, subconic, spongy; pits somewhat rounded to irregularly elongate, grayish becoming tan; ridges pallid to cream to ochraceous with age.

Stalk 6–13 long, 3–6 cm wide, massive and often columnar to folded at base; pale to cream.

Asci 225–235 × 18–22 μm, cylindric, operculate, 8 spored; ascospores 20–22 × 12–14 μm, elliptical, 1 celled; spore print cream.

Scattered to clustered on ground under hardwoods at edge of woods, particularly in areas where elms have died recently, toward the end of the spring morel season.

This species is considered by many to be the best flavored of the morels. Some interpret *M. crassipes* as part of a *M. esculenta* variable species complex.

182 *Morchella deliciosa*

183 *Morchella esculenta*

Morchella deliciosa Fr.: Fr.   **(182)**   Edible, choice
"WHITE MOREL"

Apothecium 2–3 cm long, 1.0–2 cm wide, caplike, cylindric, conic with blunt apex; pits elongate, grayish to blackish within; ridges about 1 mm thick, irregularly anastomosing, lighter than pit interior, pallid to whitish.

Stalk about as long as cap, occasionally enlarged at base; lighter than cap, whitish to cream; hollow.

Asci 200 × 12–15 μm operculate, cylindric, 8 spored; ascospores 20–24 × 10–12 μm, elliptical, hyaline; paraphyses numerous, lightly colored, slightly enlarged at apex; spore print light yellow.

Single or scattered on ground in woods or in grassy areas at edge of woods, early Sp.

> This delicious morel tends to appear slightly before *M. esculenta* in the spring and is known by some collectors as the "gray morel." It has been considered by some to be an early season form of *M. esculenta.*

Morchella esculenta Fr.   **(183)**   Edible, choice
"SPONGE MUSHROOM"

Apothecium 3–9 cm long, 2–5 cm wide, caplike, subglobose to elongate; pits irregularly arranged to radially elongate, grayish becoming yellowish brown surrounded by paler ridges.

Stalk 4–6 cm long, 1.5–3 cm wide, diameter usually not exceeding two-thirds that of cap, slightly larger at base; white to cream; hollow, longitudinally depressed in places, dry to granulose.

Asci 220–230 × 18–22 μm, cylindric, operculate, 8 spored; ascospores 20–25 × 12–16 μm, elliptical, smooth, hyaline; paraphyses numerous, enlarged above, faintly colored; spore print light yellow.

Single to scattered on ground under hardwoods in the vicinity of standing dead elms, in old orchards, and at the edge of woods, Sp.

> This is a favorite edible fungus and probably is more hunted than any other. Efforts to grow it commercially have recently succeeded.
>
> A later fruiting *Morchella, M. crassipes,* has been interpreted by some to be a large form of *M. esculenta. M. deliciosa* may also be considered by some to be a part of this species complex.

184 *Morchella semilibera*

185 *Peziza repanda*

Morchella semilibera de Candolle: Fr. **(184)** Edible
"HALF-FREE MOREL"

Apothecium 2–4 cm long, 1.5–3 wide, caplike, bell shaped or conic; yellowish when young to brownish or olive-brown at maturity; free from stalk for about half its length; pits longer than broad reaching diameter of 5–10 mm; ridges well developed.

Stalk 6–15 × 2–3 cm, somewhat enlarged at base (to 4 cm diameter); whitish to yellow; often irregularly furrowed at base, granulose to mealy, hollow.

Asci 250–300 × 20–25 μm, cylindric, operculate, 8 spored; ascospores 24–34 × 15–21 μm, elliptical, hyaline, smooth; paraphyses numerous, slightly clavate, hyaline to lightly colored; spore print light yellow.

Single to loosely scattered on ground under hardwoods, early Sp.

This is probably the least desirable of the morels but is considered by many to be an edible mushroom of good quality when young. Some people have an adverse reaction to this species, as indeed they may to any species. This species is sometimes placed in the genus *Mitrophora.*

Peziza badia Pers.: Merat

Apothecium to 10 cm in diameter, deeply cup shaped to more flattened to infolded on one side, sessile; exterior tan becoming dark brown with age; interior (hymenium) dark brown.

Asci to 330 × 15 μm, cylindric; ascospores 17–23 × 8–10 μm, elliptical, uniseriate, hyaline to faintly colored, sculptured eventually with rather coarse warts or short interrupted ridges; paraphyses clavate, yellowish.

Scattered or clustered on ground in deciduous woods, common, Sp., S.

Peziza repanda Pers. **(185)**

Apothecium large, up to 8–10 cm in diameter, cup shaped, shallow to expanded and flattened, sessile or very short stipitate; exterior whitish or pale fawn; interior (hymenium) pale brown becoming darker; margin even or splitting with age, regular in outline to irregularly folded.

Asci 225–300 × 12–15 μm, cylindric, operculate, 8 spored; ascospores 14–16 × 8–10 μm, elliptical, hyaline, smooth; paraphyses slender, slightly enlarged above, yellowish or brownish.

Single or clustered on rotten logs in woods, occasionally on chip piles or soil, not uncommon, S., F.

Peziza repanda appears quite variable in form at different ages.

186 *Peziza succosa*

187 *Sarcoscypha coccinea*

Peziza succosa Berk. **(186)**

Apothecium reaching a diameter of 4–5 cm, hemispheric expanding to become shallow cup shaped, regular in outline or folded, sessile; exterior light to yellowish; interior (hymenium) browner than exterior. Flesh becoming yellow when broken.

Asci 200–225 × 12–15 μm, cylindric, operculate, 8 spored; ascospores 16–20 × 8–12 μm; elliptical, uniseriate, usually containing 2 distinct oil drops, hyaline to faintly yellow becoming sculptured with coarse warts and short ridges; paraphyses slender, enlarged at apex.

Single or in small groups on damp soil in woods, S.; common in some years.

Sarcoscypha coccinea (Jacq.) Sacc. **(187)**   Inedible
"CRIMSON CUP"

Apothecium 2–6 cm in diameter, shallow to deeply cup shaped; exterior white to almost white, more or less floccose with matted hyaline hairs; interior (hymenium) scarlet to orange-red; margin initially incurved.

Stalk short to almost sessile, stout.

Asci 400–500 × 12–14 μm, cylindric; ascospores 26–40 × 10–12 μm, elliptical to cylindric with rounded ends, uniseriate, often with 2 large oil drops or with cluster of small oil drops at each end, hyaline; paraphyses slender, slightly enlarged above with numerous red granules.

Single or in very small groups, attached to buried or partially buried sticks or wood in woods, often on basswood, common, Sp., occasionally again in late F.

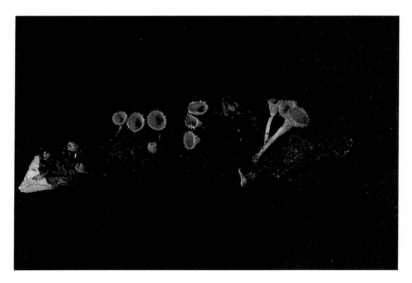

188 *Sarcoscypha floccosa*

189 *Sarcoscypha occidentalis*

Sarcoscypha floccosa (Schw.) Sacc. **(188)**   Inedible

Apothecium 5–8 mm in diameter, 1 cm deep, goblet shaped; exterior whitish covered with long, rigid, hyaline hairs reaching 1 mm in length tapered to apex giving cups a shaggy appearance; interior (hymenium) scarlet; margin strongly incurved when young; thick walled.

Stalk slender, gradually larger toward apothecium, length variable.

Asci 300–325 μm long, diameter 20 μm at apex, cylindric; ascospores 20–35 × 15–17 μm, uniseriate, elliptical, ends narrowed, smooth, hyaline or slightly yellowish; paraphyses slender, slightly thickened at tip containing numerous red granules.

Scattered to clustered, attached to buried or partially buried wood in woods, often on shagbark hickory, common, S.

This species is also known as *Microstoma floccosum* (Schw.) Rait.

Sarcoscypha occidentalis (Schw.) Sacc. **(189)**   Inedible

Apothecium reaching a diameter of 1 cm (more rarely 2–3 cm), shallow to cuplike; exterior whitish; interior (hymenium) bright red-orange to almost scarlet.

Stalk variable in length often reaching 2–3 cm and a diameter of 2 mm.

Asci cylindric; ascospores 20–22 × 10–12 μm, elliptical, uniseriate, usually with 2 oil drops, hyaline to slightly yellowish; paraphyses slender, slightly thickened above reaching diameter 3–4 μm at apex, with numerous red granules.

Scattered to clustered, attached to buried or partially buried wood, often on shagbark hickory, common, late Sp., S., F.

Stalks will vary in length depending upon how deeply the sticks colonized by the fungus are buried in litter.

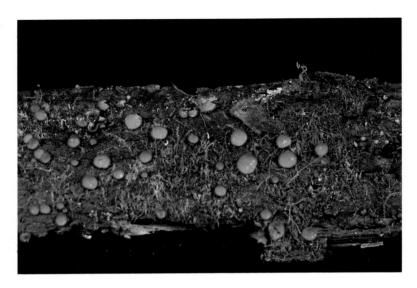

190 *Scutellinia scutellata*

191 *Spathularia flavida*

Scutellinia scutellata
(Linnaeus: St. Amans) Lambotte **(190)**

Apothecium up to 10 mm in diameter, almost globose when young becoming flattened; exterior appearing dark brown because of brown, thick-walled hairs with forked bases tapering to apex and extending to 1 mm beyond margin; interior (hymenium) bright red to orange-red.

Asci 200–225 × 12–15 μm, cylindric; ascospores 20–24 × 12–15 μm, elliptical, uniseriate, hyaline, walls with minute warts; paraphyses 18–19 × 10–12 μm, strongly enlarged above with diameter of 7–10 μm at apex, containing numerous red granules, green with iodine solution.

Clustered to crowded on well-rotted wood, often among mosses, more rarely on wet soil, very common, Sp., S., F.

Spathularia flavida Fr. **(191)**

Fruiting bodies up to 5 cm long, apical portion (modified apothecium) flattened, fan shaped to roundish; yellowish when mature, becoming darker with age; edge even or undulating, decurrent along stalk from one-third to one-half the total length.

Stalk hollow, slender, paler than apical portion.

Asci 100–125 m × 12–14 μm, clavate, inoperculate, 8 spored; ascospores 40–50 × 2–3 μm, clavate-filiform becoming multiseptate, hyaline; paraphyses filiform, branched, hyaline, curled or coiled at apex.

Single to gregarious on soil or in humus in coniferous woods, usually under pines, not uncommon, S.

192 *Urnula craterium*

193 *Verpa bohemica*

Urnula craterium (Schw.: Fr.) Fries **(192)**

Apothecium 3–4 cm in diameter, 4–6 cm deep, at first closed but opening soon leaving a ragged or smooth inrolled margin around a star-shaped to round opening, tapers gradually to a stalk; exterior brownish black to black, somewhat velvety; interior (hymenium) brownish black, slightly paler than the outside; hairs of outer wall variable in length, thick walled, flexuous, blunt.

Stalk to 3–4 cm long with mass of black mycelium at base.

Asci 600 × 15–17 μm, cylindric, operculate; ascospores 25–35 × 12–14 μm, broad, elliptical, uniseriate, smooth, hyaline; paraphyses filiform, slightly enlarged above, pale brown.

Single or clustered, attached to buried sticks (oak) on ground in deciduous woods, common but not conspicuous, Sp.

Verpa bohemica (Krombh.) Schroet. **(193)**   Not recommended

Apothecium 2–4 cm in diameter, 2–5 cm long, conic to bell shaped, folded into longitudinal ridges that anastomose frequently; yellow-brown to brown to reddish brown; attached at top only.

Stalk 6–12 cm long, 1–2.5 cm wide tapering upward; white to cream; loosely stuffed becoming hollow, smooth.

Asci 250–350 × 25 μm, cylindric, operculate, 2 spored; ascospores 60–80 × 15–18 μm, elliptical, smooth, hyaline to yellowish; paraphyses enlarged to 7–8 mm in diameter at apex, numerous; spore print yellow.

Single to scattered on ground in woods. In Iowa collected only in northeastern area, very early Sp.

This fungus has been eaten, but it **may** cause gastrointestinal upset and lack of muscular coordination. Illness seems to occur usually when larger quantities are eaten. It cannot be recommended.

Verpa conica (Mull.: Fr.) Swartz **(194)**

Apothecium 2 cm long, 1–3 cm wide, caplike, subconic to bell shaped; brown with white underneath; smooth or with faint furrows near margin, attached at top only, flaring at lower edges with age.

Stalk 5–11 cm long, 1–1.5 cm wide tapering upward; white to cream; loosely stuffed becoming hollow.

Asci to 350 × 23 μm, cylindric, operculate, 8 spored; ascospores 20–26 × 12–16 μm, elliptical, smooth, hyaline; paraphyses numerous, gradually clavate; spore print light yellow.

On ground in wooded areas, Sp.

This species is not common in all years.

194 *Verpa conica*

Other Ascomycetes

The cup fungi make up a relatively small segment of the ascomycetes. The others, the majority of ascomycetes, produce small, roundish to flask-shaped fruiting bodies. They are much less obvious and are not noticed often except by those who look carefully. Many of these fungi live on dead plant debris and produce scattered fruiting bodies on or in these materials. The fruiting bodies may be black dots on rotting down wood, dead branches in trees, annual plant stems of the preceding year, or even the surface of living leaves. The fruiting bodies may also be red or yellowish, clustered roundish structures on tree bark or dead branches, or on the larger fleshy fruiting bodies of other fungi.

Some of these Ascomycetes will be on dead plant materials, decomposing them and recycling their components. Others will be plant parasites of diverse kinds, some of them forming fruiting bodies only on overwintered debris.

In this field guide we have included the very common members of this group of Ascomycetes that produce large macroscopic structures on or in which the much smaller fruiting bodies develop, usually many in number and grouped in various ways. These larger structures, called stromata if they remain attached to the material in which vegetative mycelium developed and sclerotia if the macroscopic structure separates from where it was developed, have visible characteristic features. The following key is based upon these observable field details.

KEY TO THE GENERA AND SPECIES OF OTHER ASCOMYCETES IN THIS GUIDE

1a. Stromata flat, rounded, or clublike; on down wood, stumps, at the base of standing trees, or over larger fungus fruiting bodies such as mushrooms, boletes, and brackets 2
 b. Stromata somewhat fleshy, clublike; growing from insects in soil or rotting wood, often parasitic on the juvenile stages, or from truffles buried in the soil . . . .*Cordyceps* species

2a. Stromata fingerlike structures, tough to hard, upright, gray to black; single or in clusters on rotting wood or at the base of trees . . . .
 . . . .*Xylaria polymorpha*
 b. Stromata flat or hemispherical structures; on wood or on large fungus fruiting bodies. 3

3a. Stromata flat; on wood or on large fungus fruiting bodies 4
 b. Stromata hemispherical; on wood 6

4a. Stromata bright orange or reddish purple in age, parasitic on mushrooms; mushroom gills absent or poorly developed. . . . .
. . . .*Hypomyces lactifluorum*
  b. Stromata on wood; variously colored. 5

5a. Stromata soft, bright yellow, not limited in size. . . . .
. . . .*Hypocrea sulphurea*
  b. Stromata hard, black, when young often covered with a gray, green, or red-purple powdery layer. . . . .*Hypoxylon* spp.

6a. Stromata with concentric zones (easily visible when stromata are cut or broken), with a single outer layer of flask-shaped fruiting bodies . . . .
. . . .*Daldinia concentrica*
  b. Stromata lacking zonation, similar texture throughout with a single outer layer of flask-shaped fruiting bodies . . . .*Hypoxylon* spp.

Cordyceps militaris (L.: Fr.) Lk.   (195A)

Fruiting structure (stroma) up to 5 cm long, 3–5 mm wide, club shaped; orange to orange-red; surface of upper portion raised around each flask-shaped perithecium.

Stipe narrower in width than the club-shaped upper portion, varying in length depending on depth the parasitized insect host is buried in soil or litter.

Perithecia buried in stroma in a single layer all over upper portion of stroma.

Asci cylindrical, in inner layer of perithecia, 8 spored; ascospores long and threadlike, often breaking into segments 3.5–6 × 1.5 mm when mature.

Usually solitary or in groups of a few arising from buried parasitized Lepidoptera larvae and pupae, common in woods, S. or F.

The club-shaped stromata are easily confused with unbranched coral fungi. Attachment to a parasitized insect can be determined by carefully digging around the base of the fruiting body.

Several other species of *Cordyceps* are not uncommon in woods in summer and fall. *Cordyceps melolonthae* var. *melolonthae* parasitizes Coleoptera larvae, and is most common on large larvae of June bugs. It produces yellow clavate heads up to 13 cm long, 5–15 mm wide.

Two other common *Cordyceps* species parasitize buried fruiting bodies of *Elaphomyces*, one of the truffle fungi: *Cordyceps ophioglossoides* (195B) produces reddish to olive-brown clavate stromata up to 8 cm long in midsummer woods; *Cordyceps capitata*, a fall species, produces a stalk up to 11 cm long, capped by a rounded head in which the perithecia are buried.

A

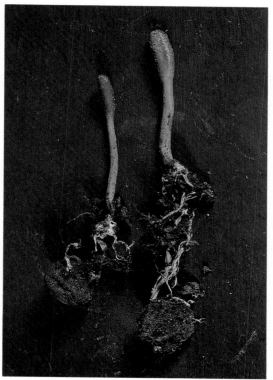

B

195 Parasitic Ascomycetes. **A:** *Cordyceps militaris* on pupae; **B:** *Cordyceps ophioglossoides* on truffles

196 *Daldinia concentrica*

197 *Hypocrea sulphurea*

Daldinia concentrica (Bolton: Fr.) Cesati & de Not. **(196)**

Stroma hemispherical; black covered with a purple-brown to reddish brown layer in young stages; hard, smooth; internal concentric zones of hard whitish and black carbonous tissue, even when young.

Perithecia buried in the outer surface in a single layer, flask shaped with openings just breaking the outer surface of stroma.

Asci 200 × 12 μm, cylindrical, 8 spored; apical plug staining blue with iodine solution or Melzer's reagent; ascospores elliptical with one flattened side, 1 celled, in a single row, dark brown to black.

On wood or bark of dead trees, usually oaks or ironwood, very common, mid-S., F.

Stroma and surrounding wood often covered by black spores as the mature ascospores are violently released from the asci. Field identification is very easy as the concentrically zoned, hard black stroma is unique.

Hypocrea sulphurea (Schw.) Sacc. **(197)**

Fruiting structure (stroma) broad, often at least several centimeters in diameter, flat, irregular in outline, effuse without a definitive size; white when young but quickly becoming bright yellow with an irregular white edge; firm but fleshy.

Perithecia 150–250 μm in diameter, oval, completely buried in the stroma and evident only as water-soaked dots on the surface of the stroma, flask shaped with the opening at the surface of the stroma, containing a layer of asci.

Asci 75–150 × 5–6 μm, cylindrical with an apical pore, 8 spored; ascospores, 3–4 × 9–11 μm, 2 celled in a single row, hyaline, cells commonly separating when mature.

Common on down wood in the woods, S., F.

The conspicuous bright yellow stroma is often associated with or developed over the old, partly disintegrated fruiting bodies of jelly fungi, particularly species of *Exidia*. Dried stromata may persist for some time, especially in the fall.

198 *Hypomyces lactifluorum*

199 *Hypoxylon mediterraneum*

Hypomyces lactifluorum (Schw.: Fr.) Tul. **(198)**

Not recommended

Fruiting structure (stroma) parasitic on mushrooms, probably on species of *Russula* and *Lactarius,* which are modified or completely changed in shape by the parasite; bright orange to orange-red becoming purple with age, entirely covering mushroom.

Perithecia buried in a single layer all over the stromatized host, flask shaped with openings appearing as darker, slightly raised bumps from the general smooth surface of the orange stroma.

Asci 150–175 × 7–8 μm, cylindrical, 8 spored; ascospores 35–40 × 7–8 μm, fusiform, sometimes slightly curved, 2 celled, hyaline, rough.

On the ground in woods, not uncommon, mid-S., early F.

These bright orange parasitized mushrooms are distinctive. The stroma-covered mushroom may have the general mushroom cap and stalk shape with gills modified to blunt ridges, or it may be modified to an unrecognizable lump of mushroom tissue covered by the stroma. They will occur in the same area in the woods in successive years, failing to develop only in very dry seasons.

Other species of *Hypomyces* occur as parasites on mushrooms, boletes, jelly fungi, saddle and cap fungi, and on old fruiting bodies of various bracket fungi. Some species have an orange stroma, while others are shades of red, yellow, cream, or white.

Hypoxylon mediterraneum (de Not.) Mill. **(199)**

Fruiting structure (stroma) 2–5 cm long, 1–1.5 mm thick, varying from flat to slightly raised to somewhat rounded to elliptical, erumpent from bark; black and shining; evenly dotted with the openings of the perithecia buried in the stroma.

Perithecia 0.5–1 mm long, 0.4–0.5 mm wide, elongated flask shape, compressed when crowded; buried in stroma with only the openings (ostioles) emergent.

Asci 120–185 × 8–11 μm, cylindrical with short stalk, 8 spored; ascospores 16–23 × 6–10 μm, oblong, elliptical, ends rounded, 1 celled in a single row, dark brown; paraphyses unbranched, numerous.

On living or dead *oak* branches, occasionally on other hardwoods, usually following injury, persists throughout the year.

There are a number of common species of *Hypoxylon* that appear as a flat crust 2–10 mm wide, 5–20 mm long to a hemispherical cushion 1–8 mm in diameter. The apical plug of the asci stains blue with I_2KI.

Xylaria polymorpha (Persoon: Merat) Greville   **(200)**

"DEAD MAN'S FINGERS"

Fruiting bodies (stroma) up to 8 cm long, 2.5 cm wide, more or less club shaped, sometimes irregular or slightly lobed above; young stromata appear powdery, buff to white when covered with a layer of conidiophores and conidia (these weather away and are not replaced as the perithecia begin to develop); thin black crust on outside penetrated by flask-shaped perithecia; interior white and tough.

Stalk short, cylindrical.

Perithecia in a single layer in the stroma, flask shaped, rather regularly spaced over upper portion.

Asci 200 × 10 μm, cylindrical, 8 spored; ascospores 20–32 × 5–9 μm, tapered at both ends but with one side flattened, 1 celled in a row, dark brown to black; paraphyses filiform.

Solitary or clustered on wood or stumps at soil level, common, S., F., persisting for months.

Several other species of *Xylaria* with smaller stromata are common on down wood. *Xylosphaeria* has also been used as a name for this genus.

200 *Xylaria polymorpha*

Plant Parasitic Fungi

A very few of the mushroom fungi cause diseases of other living organisms; they usually take their food from dead plant materials or are part of a mycorrhizal association. There are many other fungi, however, that do obtain their food from living plants, resulting in serious plant diseases.

We have included four common and obvious plant parasitic fungi. These affect their hosts in such ways that their presence is evident. Many plant parasitic fungi are less obvious in their interactions, but that is a separate fascinating world beyond the scope of this book.

Apiosporina morbosa (Fr.) von Arx **(201)**

"BLACK KNOT"

Swollen, convoluted, hard, black elongate structures 1–3 cm in diameter, up to 30 cm long, develop along a twig or branch of the host plant. It begins with an expanding swelling on twig or branch, resulting in longitudinal breaks in the bark. Diseased tissues are straw yellow to brown, becoming covered with a hard black rind in which the fruiting bodies develop. Young knots may have an olive-green velvety covering that is lost as the black outer region matures. Diseased twigs often are bent at odd angles; the portion of the twig or branch beyond the knot dies once the branch has been girdled.

Many tiny fruiting bodies develop in a single layer along and around the black diseased branch. Each fruiting body contains asci in which 2-celled, hyaline ascospores develop.

Several woody species in the rose family are hosts; choke cherry, *Prunus virginiana,* is particularly susceptible among the native hosts and is very noticeable when it is infected.

201 *Apiosporina morbosa*

202 *Claviceps purpurea*

Claviceps purpurea (Fr.) Tul.   **(202)**   Poisonous
"ERGOT"

Ergot sclerotia are common in the individual flowers of inflorescences of open-pollinated grasses such as bromegrass or rye. The sclerotia are up to 3 cm long, cylindrical, rounded at the base, tapered at the apex; dark brown to purple-black; hard, separate readily from the flowering head; interior is white, hard, surrounded by a narrow black rind.

Any mature planting of rye, such as is found on erodable banks along road construction sites, contains mature plants that have the long (2–3 times as long as the rye grains), dark sclerotia.

Young developing sclerotia produce a secretion, "honey dew," that is attractive to some insects and sticky when touched. If mature sclerotia overwinter on the ground, one or several long-stalked stromata with rounded heads about 2 mm in diameter develop in the spring. Perithecia form in a single layer on the head.

Ergot sclerotia have probably been responsible for more poisoning deaths than any other fungus. All sclerotia must be removed from any grain to be used as human or animal food. Even wheat may occasionally contain sclerotia, but much less frequently than does rye.

203 Stages of cedar-apple rust. **A:** Gall of *Gymnosporangium juniperi-virginianae* on juniper; **B:** Fruiting bodies of *G. globosum* on hawthorn

Gymnosporangium
juniperi-virginianae Schw. (203A, B)
"CEDAR-APPLE RUST"

This parasitic fungus stimulates host cell growth producing solid brown galls on the twigs or smaller branches of juniper, *Juniperus virginiana*. The galls range to several centimeters in diameter and are most obvious in rainy weather in April and May when each gall is covered by orange gelatinous spore masses **(A)**. As these gelatinous masses dry, millions of wind-borne spores are released. The spore masses may swell and dry a number of times during the spring. The spores cannot grow on juniper but can only germinate and grow successfully on apple leaves or young fruits that are just developing.

By July or August, the fungus produces clusters of very small, cuplike fruiting bodies with masses of dry orange spores on the apple leaves or fruits **(B)**. These spores in turn cannot grow on apple but must be carried by the wind back to a juniper, where they germinate and become established in a twig or small branch. The next growing season a gall develops as the fungus stimulates the juniper cells; by September the gall is mature. When rain falls the following spring, orange gelatinous masses of spores will be produced on the gall. A spore-producing gall functions for a single spring and then dies, as does the twig portion beyond the gall.

Another species, *Gymnosporangium globosum*, also produces galls on juniper, but hawthorne is the alternate host. After a wet July, the hawthorne leaves may seem to be completely spotted with clusters of orange spore-producing cups of this rust.

While many rusts share the peculiar situation of alternate development on two very different host plant species, some trudge through a similar cycle on a single host plant species; there are variations on the basic scheme, in confusing array. Most flowering plants are hosts to at least one rust species. Grasses particularly may have several rust species sharing a single host plant.

Ustilago maydis (DC.) Cda.   **(204)**   Edible
"CORN SMUT"

Small to large galls are produced from young diseased tissue in various parts of the corn plant, most commonly in the flowers of the tassel or the ear. Young galls are firm and light green but soon become soggy and dark as the fungus cells in the gall become modified as individual cells with dark thick walls. At these young stages, "corn smut" galls are eaten by some people and valued as food. Finally, the outer gall tissues break and a dry powdery spore mass is exposed.

The dry spores blow about or remain on plant tissue in the field over winter. When young corn tissues are exposed the following season, the surviving spores germinate to form thin-walled spores that enter the young corn plant cells; this can happen at any time during the growing season.

Many grasses and other plants develop similar dark smut spore masses in the flowers or leaves. Each smut fungus can develop in a specific group of plant hosts, or sometimes only in a single species of plant. The smuts that develop in flowers usually replace the fruits or seeds of the normal plant. In such plants, they are dangerously successful competitors with humans for the food produced by the green plant host.

204 *Ustilago maydis*

Myxomycetes
(Slime Molds)

THE myxomycetes, or slime molds, are common organisms but only a few produce large and relatively conspicuous fruiting structures. Their vegetative stages, called plasmodia, occur as naked protoplasm dispersed in well-rotted logs, leaf mulch, or piles of straw or similar decaying plant material (205). Their fruiting structures occur on the outside of rotten logs, along living plant stems or leaves, or on the upper leaves in leaf litter on the woodland floor. Many of these fruiting bodies, called sporangia or aethalia, are quite small (less than a millimeter in height) and inconspicuous; some are larger and more obvious or are grouped together in large associations and thus are more conspicuous. All of them contain a dry powdery spore mass, which readily blows about when the outer covering of the fruiting structure is broken.

Slime mold fruiting bodies may develop at any time during the growing season but are more common in the summer and fall. They frequently develop within a week or so of dry weather following a rain or a rainy period as the substrate in which the plasmodia have been feeding dries. Freshly developed fruiting bodies are most striking and the easiest to recognize. However, they are rather fragile, and the spores are washed away by rains, leaving the fruiting bodies drab and unrecognizable.

Only a few of the common slime molds with the most conspicuous fruiting bodies are discussed here. There are several books devoted to slime mold information and identification. Those most useful in this area are listed in the Technical References.

205   Conspicuous plasmodium of a slime mold

KEY TO GENERA IN THIS GUIDE

1a. Fruiting bodies usually larger than 0.5 cm in diameter; on a broad base 2
  b. Fruiting bodies smaller than 0.5 cm in diameter; often with obvious stalks. 4

2a. Fruiting bodies a black powdery spore mass covered with a white to yellowish flaky layer . . . .*Mucilago* and *Fuligo* (Species of these genera look alike macroscopically but can be easily recognized by using microscopic differences)
  b. Fruiting bodies with a light-colored spore mass covered by a smooth, somewhat persistent thin layer; on wood. 3

3a. Fruiting bodies globose to flattened, bright pink when young becoming gray to tan in age; powdery spore mass. . . . .*Lycogola*
  b. Fruiting bodies with a somewhat constricted base, bright red when young becoming reddish brown in age; powdery spore mass reddish brown. . . . .*Tubifera*

4a. Spore mass brightly colored, often golden yellow; outer covering of fruiting bodies soon breaking exposing powdery spore mass and threadlike capillitium. . . . .*Trichia* and *Hemitrichia*
  b. Spore mass brown to black 5

5a. Fruiting bodies in a close cluster; each fruiting body with a wiry central stalk, resembling a tiny cattail . . . .*Stemonitis*
  b. Fruiting bodies in a common area, but usually distinct and separate; round to somewhat elongate with or without a short stalk 6

6a. Fruiting body goblet shaped with a persistent wall and a definite lidlike area; if lid has broken away white lumps in the exposed dark spore mass visible. . . . .*Craterium*
  b. Fruiting bodies spherical to slightly flattened; wall readily breaking into pieces and falling off or wall appearing as a dry white powdery covering 7

7a. Fruiting bodies with a dry white powdery covering . . . . . .*Didymium*
  b. Fruiting bodies with outer wall readily breaking into irregular pieces and falling away . . . .*Physarum*

206   *Craterium aureum*

207   *Didymium* sp.

Craterium aureum (Schum.) Rost. **(206)**

Sporangia about 0.4–0.6 mm in diameter, 0.7–1.5 mm high, ellipsoidal to goblet shaped; golden yellow to greenish yellow; apical portion of peridium breaks away leaving an opening that exposes spore mass.

Stalk short.

Spore mass dark; capillitium with scattered light-colored lime knots.

Occurs in large groups on old leaves of ground litter in woods, S., F.

C. leucocephalum has slightly larger sporangia and is the most common *Craterium* species.

Didymium **(207)**

Sporangia small, usually 1 mm or less in height and diameter; white; walls of commonest species with lime crystals giving tiny sporangia appearance of having been dusted with powdered sugar.

Stalk present or absent (sessile).

Spore mass uniformly dark; capillitium nonlimy.

Common in large numbers on old leaves of ground litter in woody areas, S., F.

Species of *Didymium* are very similar to those of the genus *Physarum*.

208 *Fuligo septica*

209 *Lycogola epidendrum*

Fuligo septica (L.) Wiggers **(208)**

Fruiting bodies (aethalia) large, 2–20 cm broad, 1–3 cm high, pulvinate; white, thick, flaky cover.

Spores 6–9 μm in diameter, round, 1 celled, black; spore mass dull black, powdery; capillitium very fine threads, uncolored, white deposit areas (lime knots) randomly scattered along threads throughout.

On rotten wood, plant litter, and leaves and/or stems of living plants, especially if they are surrounded by a heavy mulch where the plasmodium can grow, S. (most common), F.

Lycogola epidendrum (L.) Fr. **(209)**

Fruiting bodies (aethalia) 3–15 mm broad, hemispherical to slightly depressed; at first appearing bright pink maturing in a few hours to gray, gray-brown, or olivaceous; cover tough, rough.

Spores 6–7.5 μm, globose, 1 celled, very lightly colored with ridges (reticulations) on spore wall; spore mass dull pinkish gray to yellow-tan, randomly exposed when the cover breaks.

Always occurs on wood, S., F.

The clusters of fruiting bodies look like very small puffballs and are often incorrectly identified as such. Another slime mold, *Reticularia lycoperdon,* also forms large (up to 8 cm in diameter), hemispherical to slightly flattened, puffball-like fruiting bodies on rotting logs in woods. When the tough outer covering breaks, the dry inner spore mass is reddish brown, resembling cocoa in color and texture.

210 *Mucilago crustacea*

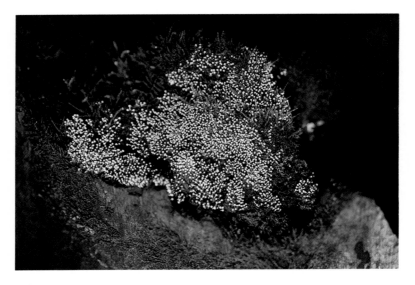

211 *Physarum* sp.

Mucilago crustacea Wiggers **(210)**

Fruiting bodies (aethalia) 1–7 cm long, 1–2 cm thick; white, creamy with a flaky cover of white calcareous crystals; often surrounded by a well-developed, thin, spongy to membranous white mat (hypothallus).

Spores 11–13 μm in diameter, round, 1 celled, densely warted, black; spore mass dark black mixed with very thin branched threads (capillitium).

On logs or bases of trees in woods, often on grasses in lawns or pastures, common on mulch, S., F.

Fuligo and *Mucilago* look quite similar and can be identified with certainty only when differences in the form of the lime deposits and the details of the capillitium are examined with a light microscope. They are the most conspicuous and thus the most frequently collected slime molds.

Physarum **(211)**

Fruiting bodies (sporangia) small, usually not more than 3 mm high, round to lobed; white or colored and may contain lime deposits; outer wall (peridium) quite thick and persistent or thin and fragile.

Stalks present or absent.

Spore mass dry, dark black to purple-brown when exposed; capillitium dotted at intervals with deposits of white lime knots, fine, threadlike, branched throughout, not easily seen.

Sporangial groups loosely scattered to closely clustered with large numbers of individual sporangia (sometimes hundreds or thousands) developed concurrently over rotted logs or on dead leaves and ground litter, S., F.

Microscopic features must be considered in distinguishing *Physarum* from other dark-spored sporangiate genera and for identifying species.

Physarum cinereum is nonstalked, white, with crowded sporangia; common on living grass leaves in lawns or on other living plant leaves in well-mulched gardens.

Physarum viride is stalked with flattened sporangia; wall cracked to resemble a ginger cookie pattern; common on well-rotted logs in wooded areas.

Physarum polycephalum produces yellowish to gray, lobed and convoluted, stalked sporangia; not uncommon on rotted logs and adjacent litter. It is well known in biology laboratories in its vegetative stage, for the rapidly growing yellow plasmodium is easily cultured and maintained.

Species of *Craterium*, *Didymium*, and *Badhamia* also are common in many of the habitats where *Physarium* species grow, all developing most frequently S., F.

212 *Stemonitis* sp.

213 *Trichia* sp.

Stemonitis **(212)**

Fruiting bodies long, cylindric, closely clustered, groups often several cm across; purple-brown to black.

Sporangia slender, often curved; upper spore-filled portions dull purple-brown to black above thin shiny black stalks.

Stalks from a few mm to over 1 cm long, continue into the sporangium as a central column with branched threadlike capillitium surrounded by powdery dark spores; this inner structure obvious as spores blow away or wash away in a rain.

Common on down logs in woods, S., F.

Stemonitis splendens is the most common species with large clusters of long (may be 2 cm or more), slender, dark purplish brown sporangia. *S. fusca* is shorter with dark brown sporangia. Species cannot be identified with certainty without use of microscopic information about the spores and capillitium.

Trichia **(213)** and *Hemitrichia*

Individual sporangia of species of *Trichia* and *Hemitrichia* usually 0.5–1 mm in diameter, roundish, often occur tightly grouped in clusters that may be several centimeters across; golden yellow.

Stalk short or absent.

Once somewhat shiny outer wall (peridium) is irregularly broken away, bright yellow powdery inner spore mass and capillitium are exposed and readily dispersed by wind and splashing rain.

Common on decayed logs, mid-S., F.

Species of *Trichia* are more common; however, identification to genus and species can be made only when microscopic examination of the capillitium is made.

Tubifera **(214)**

Sporangia often 0.5 mm in diameter, several mm in height, grouped in clusters up to 1 cm across; bright clear red when young maturing in a few hours to dull reddish brown.

Spores reddish brown, dry, quickly dispersed when thin outer cover (peridium) breaks.

Clusters of sporangia occur on well-rotted logs, a number of clusters often developing from the same plasmodium and occurring in the same area, S., F.

This slime mold is much more conspicuous in the young sporangial stage than at maturity.

214   *Tubifera* sp.

Glossary

acrid: peppery or biting in taste.

adnate: broadly attached to the stalk (gills).

adnexed: reaching to the stem, not completely attached to it (gills).

aethalium (pl. aethalia): sessile fruiting body formed from plasmodium of myxomycetes.

allantoid: sausage shaped, curved (spores).

amyloid: having a blue reaction in the presence of iodine (as in Melzer's reagent).

anastomosing: joining crosswise to form angular areas (of gills).

annulus: ring of partial veil tissue left on the stalk of a mushroom.

apical: situated toward the upper part, as the portion of stalk nearest the gills or cap.

apical pore: a differentiated area in the spore wall through which the germ tube emerges.

apiculus: a small projection on the basidiospore where it was attached to the sterigma.

apothecium (pl. apothecia): cup-shaped to saucerlike to saddle-shaped fruiting body of ascomycetes.

appendiculate: having fragments of veil hanging from cap margin.

appressed: pressed close to or lying flat against.

ascus (pl. asci): saclike cell in ascomycetes in which ascospores are formed.

basidium (pl. basidia): hyphal cell in basidiomycetes on which the basidiospores are formed.

biseriate: arranged in two rows, as ascopores in an ascus.

bolete: stalked fleshy fruiting body having a cap with tubes on the lower surface.

boletinoid: having the undersurface of the cap with pore walls developed unevenly to appear as radial rows from stalk to margin.

bracket: shelflike fruiting body produced by wood-rotting basidiomycetes.

brittle: rigid and breaking with sharp snap; shattering (cap or stalk).

bulb: enlarged base of the mushroom stalk.

campanulate: bell shaped (cap).

capillitium: a mass of threadlike strands among the spores of some puffballs and slime molds.

chlamydospore: a hyphal cell modified to be an asexual spore.

clavate: club shaped (cap).

conk: a fruiting body of a wood-inhabiting fungus, especially a polypore.

context: mycelial tissue between the surface of the fruiting body and the spore-bearing area (gills or tubes) of basidiomycetes.

corrugated: wrinkled, usually with long wavelike ridges.

cuticle: outer layer of a mushroom cap.

cystidium (pl. cystidia): sterile specialized cell occurring among the basidia in basidiomycetes.

decurrent: extending down the stalk (gills).

deliquescent: dissolving into a fluid, often inky, as in *Coprinus* (gills, etc.).

dextrinoid: staining yellow or reddish brown with Melzer's solution.

disc: central part of the cap.

discoid: flat and circular; disklike.

distant: well separated from one another (gills).

down wood: nonliving wood on the ground.

duff: partly decayed organic matter on forest floor.

eccentric: off-center (stalk attachment).

elliptical: rounded at both ends and with curved sides (spores).

erumpent: bursting through the surface of the substratum.

family: a group of genera related to one another, usually ending in "aceae."

fetid: stinking; smelling rotten.

fibril (adj. fibrillose): minute hair, usually scattered over cap or stalk.

filiform: slender; threadlike.

fleshy: of a soft, easily broken tissue (cap or stalk).

floccose: with cottony hairs.

fusiform: spindle shaped; tapered at both ends.

gall: an outgrowth produced as a result of attack by a fungus or other disease agent.

genus (pl. genera): group of related species or types of organisms; capitalized and generally italic or underlined.

glabrous: smooth; lacking hairs or scales.

glandular dots: small, moist to viscid dots on stipes, especially of some boletes.

gleba: interior tissue of a young gasteromycete fruiting body from which spores and capillitium develop.

globose: spherical; nearly round.

glutinous: slimy; very sticky.

granulose: covered by granules.

gregarious: growing together in groups (fruiting bodies).

hygrophanous: with a watery appearance when wet, and changing color when dry.

hygroscopic: capable of absorbing water readily, often swelling or changing shape.

hyaline: transparent; colorless.

hymenium: a palisadelike layer of spore-producing cells (basidia or asci).

hypha (pl. hypae): a single thread or filament of the vegetative body of a fungus.

indusium: the netlike structure hanging from the top of the stalk under the cap in some stinkhorns (*Dictyophora duplicata*).

infundibuliform: funnel shaped.

inoperculate: opening in some way other than by an operculum (lid), as in ascus of ascomycetes.

lacerate: appearing as if torn.

lamella (pl. lamellae): gill of a mushroom.

latex: milky juice found in some mushrooms.

margin: edge or outer portion of cap.

mealy: having granular texture.

Melzer's reagent: solution widely used for amyloid test (blue reaction) of fungal cell walls or spores; 20 ml water, 1.5 gm potassium iodide, 0.5 gm iodine, and 20 gm chloral hydrate.

membranous: skinlike; thin and pliant.

micrometer (micron): 1000 micrometers are equal to 1 millimeter; μm.

mushroom: a fleshy fruiting body with gills.

mycelium (adj. mycelioid): the collective network of hyphae (filaments) of the vegetative part of fungi.

mycology: the study of fungi.

mycorrhiza: the symbiotic association of a fungus and the roots of a vascular plant.

obovoid: ovoid with the broad end toward the apex.

operculate: with a lid (operculum) at the apex of the ascus.

order: a group of related families, usually ending in "ales."

ovoid: egg shaped.

paraphyses: sterile hyphae among spore-producing cells in a hymenium, usually among asci.

partial veil: a tissue cover extending from the margin of a young, unexpanded mushroom cap to the stalk.

peridiole: a portion of the gleba surrounded by a separate wall in the "Bird's Nest Fungi."

peridium: the outer wall of a fruiting body, particularly in gasteromycetes and slime molds.

pileus: the cap or expanded portion at apex of the stalk of fleshy fungi such as mushrooms.

plane: having a flattened mushroom cap, usually a mature cap.

pliant: capable of bending without breaking (stalk).

poroid: having lamellae joined by cross-veins so as to appear like pores.

pubescent: minutely hairy; downy.

pulvinate: cushionlike.

pustulate: covered with pustules or similar prominences.

pyriform: pear shaped.

resupinate: flat on the substrate with hymenium outward (fruiting body).

reticulate: having a network of ridges.

rhizomorph: a cord- or rootlike aggregation of fungal mycelium.

ring: remains of the partial veil attached to the stalk (annulus).

scabers: dark tufts of hairs, especially on the stipe of *Leccinum* in the boletes.

scale: a piece of tissue resembling a shingle on the cap or stalk.

sclerotium (pl. sclerotia): hard, dark mass of fungal tissue, usually a resting structure.

scurfy: having a localized darkened and rough surface; dry flaky surface.

septum: a cross-wall in a fungal hypha or spore.

serrate: toothed; notched like a saw.

sessile: with no stalk.

setae: pointed, thick-walled sterile cells.

sinuate: notched at the stalk (gills).

spathulate: spoon shaped or tonguelike.

species: one sort of organism; the second part of a binomial epithet; lower-case and generally italic or underlined.

sporangium (pl. sporangia): a saclike cell within which asexual spores are formed.

spore: single or multicelled reproductive structure of a fungus.

squamule: a small scale of indefinite shape.

sterigma: projection from the basidium on which a basidiospore is formed.

stipe (adj. stipitate): stalk supporting the cap of the mushroom; stemlike structure.

striate: marked with minute lines or furrows.

subdistant: fairly well separated, but less than distant (gills).

subfusiform: somewhat tapered but not spindle shaped.

subglobose: almost or nearly rounded.

subzonate: indistinctly zoned as color differences on a mushroom.

superior: located near the top of the stalk (annulus).

tomentose: densely covered by hairy or wooly growth.

truncate: ending abruptly (spores with blunt ends).

tuberculate: covered with warty or knoblike projections.

turbinate: top shaped.

umbilicate: having a central depression (cap).

umbonate: having a knob or raised swelling in the center of the cap.

uniseriate: arranged in a single row, as ascospores in an ascus.

universal veil: a layer of tissue completely surrounding a young fruiting body, such as a young mushroom.

velar: of tissue from the universal veil of a mushroom.

viscid: sticky to the touch; not deeply slimy.

volva: universal veil portion surrounding the base of the stalk or fragmenting to patches of remnant tissue around the bulb.

zonate: having concentric zones or bands of color on the cap.

General References

Alexopoulos, C. J., and C. W. Mims. 1979. Introductory Mycology, 3d ed. New York: John Wiley and Sons.

Ammirati, J. F., J. A. Traquair, and P. A. Horgan. 1985. Poisonous Mushrooms of the Northern United States and Canada. Minneapolis: Univ. of Minn. Press.

_____. 1985. Poisonous Mushrooms of Canada including Other Inedible Fungi. Minneapolis: Univ. of Minn. Press.

Anon. 1963. Kitchen Magic with Mushrooms. Berkeley: Mycol. Soc. San Francisco.

Bessette, A., and W. Sundberg. 1987. Mushrooms, a Quick Reference Guide to Mushrooms of North America. N.Y.: Collier Books.

Christensen, C. M. 1943/1974. Common Fleshy Fungi. Minneapolis: Burgess.

Glick, P. G. 1979. The Mushroom Trailguide. New York: Holt, Rinehart, and Winston.

Graham, V. O. 1944/1970. Mushrooms of the Great Lakes Region. New York: Dover Publ.

Groves, J. Walton. 1962/1979. Edible and Poisonous Mushrooms of Canada. Publication 1112, Canada Dept. of Agric., Ottawa.

Harris, R. 1976. Growing Wild Mushrooms. Berkeley, Calif.: Wingbow Press.

International Code of Botanical Nomenclature. 1983. Ed. E. G. Voss et al. Utrecht: Bohn, Sheltema & Holkema.

Kaufmann, C. H. 1918/1971. The Agaricaceae of Michigan. 2 vols. Michigan Geol. and Biol. Sur. Publ. Biol. Ser. 5, Lansing. Reprint, The Gilled Mushrooms (Agaricaceae) of Michigan and the Great Lakes Region. New York: Dover Publ.

Lange, M., and F. B. Hora. 1963. A Guide to Mushrooms and Toadstools. New York: E. P. Dutton.

Lincoff, Gary H. 1981. The Audubon Society Field Guide to North American Mushrooms. New York: Alfred A. Knopf.

Lincoff, G., and D. H. Mitchel. 1977. Toxic and Hallucinogenic Mushroom Poisoning. New York: Van Nostrand Reinhold.

McIlvaine, C., and R. K. Macadam. 1902/1973. One Thousand American Fungi. New York: Dover Publ.

McKenney, M., and D. Stuntz. 1971. The Savory Wild Mushroom. Seattle: Univ. of Wash. Press.

McKnight, K. H., and V. B. McKnight. 1987. Peterson Field Guide to Mushrooms. Boston: Houghton Mifflin.

Menser, G. P. 1977. Hallucinogenic and Poisonous Mushroom Field Guide. Berkeley, Calif.: And/or Press.

Miller, Orson K., Jr. 1979. Mushrooms of North America. 3d ed., softback. New York: E. P. Dutton.

Miller, Orson K., Jr., and Hope H. Miller. 1980. Mushrooms in Color. New York: E. P. Dutton.

Miller, Orson K., Jr., and D. F. Farr. 1975. An Index of the Common Fungi of North America. Synonymy and Common Names. W. Ger.: J. Cramer.

Pacioni, Giovanni. 1981. Simon and Schuster's Guide to Mushrooms. U.S. ed. Gary Lincoff. New York: Simon and Schuster.

Rumack, B. H., and E. Salzman. 1978. Mushroom Poisoning: Diagnosis and Treatment. West Palm Beach, Fla.: CRC Press.

Shiosaki, P., ed. 1969. Oft Told Mushroom Recipes. Seattle, Wash.: Puget Sound Mycol. Soc. Republished, Wild Mushroom Recipes. Seattle, Wash.: Pacific Search.

Smith, A. H., and Nancy S. Weber. 1980. The Mushroom Hunter's Field Guide. Ann Arbor: Univ. of Michigan Press.

Smith, A. H., H. V. Smith, and N. S. Weber. 1979. How to Know the Gilled Mushrooms. Dubuque, Iowa: Wm. C. Brown.

Smith, Helen V., and A. H. Smith. 1973. How to Know the Non-Gilled Fleshy Fungi. Dubuque, Iowa: Wm. C. Brown.

Stubbs, A. H. 1971. Wild Mushrooms of the Central Midwest. Lawrence: Univ. of Kans. Press.

Sundberg, W. J., and J. A. Richardson. 1980. Mushrooms and Other Fungi of Land Between the Lakes. Knoxville: Tenn. Val. Auth.

Thiers, H. 1975. California Mushrooms: A Field Guide to the Boletes. New York: Hafner Press.

Tracy, Marian. 1968. The Mushroom Cookbook. Garden City, N. Y.: Doubleday.

Watling, R. 1973. Identification of the Larger Fungi. London: Hulton Educ. Publ.

Technical References

Agaricales

Bas, C. 1969. Morphology and subdivision of *Amanita* and a monograph of its section *Lepidella*. Persoonia 5(4):285–579.

Bigelow, H. E. 1982. North American species of *Clitocybe*. Part I. Nova Hedw. Beih. 72:1–280.

———. 1985. North American species of *Clitocybe*. Part II. Nova Hedw. Beih. 81:281–471.

Coker, W. C. 1944. The smaller species of *Pleurotus* in North Carolina. J. Elisha Mitchell Sci. Soc. 60:71–95.

Coker, W. C., and A. H. Beers. 1943/1974. The Boleti of North Carolina. Reprint. New York: Dover Publ.

Gilliam, M. S. 1976. The genus *Marasmius* in the northeastern United States and adjacent Canada. Mycotaxon 4:1–144.

Grund, D. W., and D. F. Stuntz. 1977. Nova Scotian Inocybes. IV. Mycologia 69:392–408.

Halling, R. E. 1983. The genus *Collybia* (Agaricales) in the northeastern United States and adjacent Canada. Mycol. Mem. 8:1–148.

Hesler, L. R. 1967. Entoloma in southeastern North America. Nova Hedw. Beih. 23:1–196.

Hesler, L. R., and A. H. Smith. 1963. North American Species of *Hygrophorus*. Knoxville: Univ. of Tenn. Press.

———. 1965. North American Species of *Crepidotus*. Darien, Conn.: Hafner.

———. 1979. The North American Species of *Lactarius*. Ann Arbor: Univ. of Mich. Press.

Homola, R. L. 1972. Section Celluloderma of the genus *Pluteus* in North America. Mycologia 64:1211–47.

Jenkins, D. 1977. A taxonomic and nomenclatorial study of the genus *Amanita* section *Amanita* in North America. Bibl. Mycol. 57:1–126.

Miller, O. K. 1971. The genus *Gomphidius* with a revised description of the Gomphidiaceae and a key to the genera. Mycologia 63:1129–63.

Shaffer, R. L. 1957. *Volvariella* in North America. Mycologia 49:545–79.

Singer, R. 1949. The genus *Gomphidius* Fries in North America. Mycologia 41:462–89.

Smith, A. H. 1947. North America Species of *Mycena*. Ann Arbor: Univ. of Mich. Press.

———. 1951. North American species of *Naematoloma*. Mycologia 43:467–521.

———. 1960. *Tricholomopsis* (Agaricales) in the western hemisphere. Brittonia 12:41–70.

———. 1972. The North American Species of *Psathyrella*. New York: Memoirs New York Botanical Garden 24.

Smith, A. H., and L. R. Hesler. 1968. The North American Species of *Pholiota*. Monticello, N.Y.: Hafner.

Smith, A. H., and Harry D. Thiers. 1971. The Boletes of Michigan. Ann Arbor: Univ. of Mich. Press.

Smith, H. V. 1954. A revision of Michigan species of *Lepiota*. Lloydia 17:307–28.

Van de Bogart, F. 1976. The genus *Coprinus* in western North America. Part I: Section *Coprinus*. Mycotaxon 4:233–75.

van Waveren, E. Kits. 1970. The genus *Conocybe* subgenus *Pholiotina*. I. Persoonia 6:119–65.

Watling, R. 1971. The genus *Conocybe* subgenus *Pholiotina*. II. Persoonia 6:313–39.

Aphyllophorales

Bigelow, H. E. 1978. The cantharelloid fungi of New England and adjacent areas. Mycologia 70:707–56.

Coker, W. C., and A. H. Beers. 1951. The stipitate hydnums of the eastern United States. Chapel Hill: Univ. of N.C. Press.

Coker, W. C., and J. N. Couch. 1923. The *Clavarias* of the United States and Canada. Chapel Hill: Univ. of N.C. Press.

Cooke, W. B. 1961. The genus *Schizophyllum*. Mycologia 53:575–99.

Dodd, J. L. 1972. The genus *Clavicorona*. Mycologia 64:737–73.

Gilbertson, R. L., and L. Ryvarden. 1986. North American Polypores. Vol. I: Abortiporus-Lindtneria. Oslo, Nor.: Fungiflora.

———. 1987. North American Polypores. Vol. 2, Megasporoporia—Wrightoporia. Oslo, Nor.: Fungiflora.

Harrison, K. A. 1961. The stipitate hydnums of Nova Scotia. Publication 1099, Dep. of Agric., Ottawa, Can.

———. 1973. The genus *Hericium* in North America. Mich. Bot. 12:177–94.

Overholts, L. O. 1953. The Polyporaceae of the United States, Alaska, and Canada. Ann Arbor: Univ. of Mich. Press.

Peterson, R. H. 1974. Contribution toward a monograph of *Ramaria* I, some classic species redescribed. Am. J. Bot. 61(7):739–48.

Smith, A. H., and E. Morse. 1947. The genus *Cantharellus* in the western United States. Mycologia 39:497–535.

Wells, V. L., and P. E. Kempton. 1968. A preliminary study of *Clavaridelphus* in North America. Mich. Bot. 7:35–57.

Gasteromycetes

Brodie, H. J. 1975. The Bird's Nest Fungi. Toronto, Ont.: Univ. of Toronto Press.

Coker, W. C., and J. N. Couch. 1928/1974. The Gasteromycetes of the Eastern United States and Canada. Reprint. New York: Dover Publ.

Smith, A. H. 1951. Puffballs and Their Allies in Michigan. Ann Arbor: Univ. of Mich. Press.

Zeller, S. M., and A. H. Smith. 1964. The genus *Calvatia* in North America. Lloydia 27:148–86.

Jelly Fungi

Martin, G. W. 1952. Revision of the North Central Tremellales. State Univ. of Iowa Stud. Nat. Hist. 19(3):1–122.

Ascomycetes

Dennis, R. W. G. 1978. British Ascomycetes. Vaduz, Liecht.: J. Cramer.

Seaver, F. J. 1928. The North American Cup-Fungi (Operculates). New York: author.

_____. 1942. The North American Cup-Fungi (Operculates) Supplement. New York: author.

_____. 1951. The North American Cup-Fungi (Inoperculates). New York: author.

Myxomycetes

Farr, M. A. 1981. How to Know the True Slime Molds. Dubuque, Iowa: Wm. C. Brown.

Martin, G. W., and C. J. Alexopoulos. 1960. The Myxomycetes. Iowa City: Univ. of Iowa Press.

Index